Henry Wylde

The Evolution of the Beautiful in Sound

A treatise, in two sections

Henry Wylde

The Evolution of the Beautiful in Sound
A treatise, in two sections

ISBN/EAN: 9783337254650

Printed in Europe, USA, Canada, Australia, Japan

Cover: Foto ©berggeist007 / pixelio.de

More available books at **www.hansebooks.com**

THE ~~EVOLUTION~~

OF THE

BEAUTIFUL IN SOUND.

A TREATISE, IN TWO SECTIONS.

Tracing up the Origin, History, and Gradual Evolution of the Modern Series of Musical Sounds, from the most ancient periods through the Greek, Ecclesiastical, and Mediæval ages to the present time.

BY

HENRY WYLDE, Mus. Doc.,

TRINITY COLLEGE, CAMBRIDGE.

The Gresham Professor of Music; Principal of the London Academy of Music; Founder and Conductor of the New Philharmonic Concerts, and Author of many Treatises on Ancient and Modern Music, &c., &c.

JOHN HEYWOOD,
Deansgate and Ridgefield, Manchester;
11, Paternoster Buildings,
LONDON.
1888.

TO THE

REV. SIR **FREDK. A. GORE OUSELEY**, BART., M.A., MUS. DOC.,

PROFESSOR OF MUSIC TO THE UNIVERSITY OF OXFORD,

IN ACKNOWLEDGMENT OF THE BENEFICIAL INFLUENCE HE HAS

EXERCISED OVER MUSIC AND MUSICIANS,

THIS WORK IS DEDICATED BY

THE AUTHOR.

DECEMBER 6th, 1888.

TABLE OF CONTENTS.

———0———

SECTION I.

SECTION II.

PREFATORY.

In this treatise it has been deemed expedient in alluding to the most ancient series of sounds (*i.e.*, the Pentatonic), to show the width and relationship of intervals by giving the lengths of the elastic bodies or pipes which generated the sounds.

The more general plan has been to describe the width or space existing between certain intervals by the "ratios" of the widths or spaces.

Pythagoras is credited with the introduction of the ratio system into Greece, but it does not appear to have been adopted by the best known Grecian writers on musical science. Their methods, as well as those of the earlier Greek periods, consisted in giving the respective fractional parts of the string which produced the intervals. As the figures in the ratio system are the same as those which show the fractional parts of a string, no mistake can arise by adopting either the one or the other system when describing ancient methods.

Aristoxenes' plan of dividing the tetrachord, or $\frac{3}{4}$ of a string, into 30 parts, and describing certain intervals by so many parts, will be found elucidated by subjoined equivalent ratios.

Euclid's method—namely, that of dividing a string into four parts, and then adding to and subtracting from the several parts—is exemplified by ratios, as also is Ptolemy's plan of showing the width of intervals by figures ranging from 60 to 120.

A description of intervals by an actual or relative number of vibrations is adopted in reference to the Pentatonic series, in order to favour comparisons with modern observations on the Javanese Salendro scale.

Ratios are the same whether used in reference to lengths or vibrations, and Pythagoras is supposed to have been aware of this, although no mention is made of vibrations in relation to his musical theories.

In connection with the method of indicating intervals by ratios, the plan is adopted of representing them by a relative number of "cents." It may be here observed, however, that there is no classical authority for this plan. The system of describing intervals by "cents" is the ingenious device of Mr. Ellis, F.R.S., and consists in dividing the octave into 1,200 parts. The intervals between notes of different pitch are then indicated by a given number of cents.

The cent system has the advantage of indicating the width of intervals more readily than that of ratios.

Both methods are used in different portions of the treatise, and the cent system is often added to that of ratios, with a view of facilitating the calculations by which intervals are described in the diagrams.

HOW TO FIND THE NUMBER OF CENTS IN A GIVEN INTERVAL.

1. If of the two numbers expressing the interval ratio, three times the larger is not greater than four times the smaller, multiply 3,477 by the difference, and divide by their sum to the nearest whole number, adding 1 to the result if over 450. Thus, if the ratio is 4 : 5, where three times the larger = 15, is less than four times the smaller number = 16, the difference is 1 and sum 9, and dividing 3477 by 9 the result is 386, the cents required.

2. If the ratio is greater than 3 : 4 and less than 2 : 3, multiply the larger number by 3 and the smaller by 4. Proceed as before, and finally add 498 to the result. Thus, for 32 : 45 multiply 45 by 3 and 32 by 4 giving 128 : 135, difference 7, sum 263. Then $7 \times 3477 \div 263$ gives 92, and $92 + 498$ gives 590, the cents required.

3. If the ratio exceeds 2 : 3 multiply the larger number by 2, and the smaller by 3, and proceed as in the first case, adding 702 to the result. Thus, for 5 : 8 take 3×5, 2×8 or 15 : 16, difference 1, sum 31 ; then, $3477 \div 31 = 112$, and this added to 702 gives 814, the required number of cents. (Ellis's contribution to the Journal of the Society of Arts, March 27th, 1885.)

N.B. The above methods do not always give the exact cents. Thus, the exact cents for ratio 8·9 are 203·910, but the errors are of no consequence for ordinary purposes.

TABLE OF CERTAIN INTERVALS EXPRESSED IN CENTS

Name	Ratios.	Cents.
Quarter Tone	239:246	50
Small Semitone	24:25	70
Pythagorean Limma	243:256	90
Small Limma	178:256	92
Equal Semitone	84:89	100
Diatonic Semitone	15:16	112
Minor Second	9:10	182
Major Second	8:9	204
Septimal Minor Third	6:7	267
Just Minor Third	5:6	316
Just Major Third	4:5	386
Pythagorean Major Third	64:81	408
Just Fourth	3:4	498
Just Fifth	2:3	702
Just Minor Sixth	5:8	814
Just Major Sixth	3:5	884
Pythagorean Major Sixth	16:27	906
Septimal or Harmonic Minor Seventh	4:7	969
Just Minor Seventh	9:16	996
Just Major Seventh	8:15	1088
Pythagorean Major Seventh	128:243	1110
Octave...	1:2	1200

" He gave man speech, and speech created thought,
Which is the measure of the universe :
And Science struck the thrones of earth and heaven,
Which shook and fell not ; and the harmonious mind
Poured itself forth in all-prophetic song :
And Music lifted up the listening spirit
Until it walked, exempt from mortal care,
Godlike, o'er the clear billows of sweet sound."

—*Shelley.*

EVOLUTION OF THE BEAUTIFUL IN SOUND.

SECTION I.

CHAPTER I.

ON THE EVOLUTION OF MUSICAL SOUNDS.

THE love of the beautiful has ever been recognised as one of the most universal attributes of the human mind.

The earnest desire to arrive at, and the clear mental perception of, that which is most beautiful in art and nature is a characteristic trait of intellectual power, and commonly prevails amongst the most highly-cultured individuals of civilised nations. The analytical understanding of the beautiful, however, or a scientific realisation of the principles which underlie true beauty, has been to a great extent a sealed book even to the philosophers of past ages, and it is only in our own decades of time that science has succeeded in removing the beautiful from the realms of mere ideality and registering it among the operations of natural law.

Without attempting in this treatise to show how invariably certain universal principles find their

B

analogies throughout every department of art and nature, it will be enough to allege that they exist; and whilst they prevail in every form by which Nature herself is systematised into scientific order, we claim that one of the most striking illustrations of these universal analogies will be found by researches into the underlying causes of the beautiful in sound.

One of the marked results of modern intellectual development is the discovery that the fine arts do not impress the mind with pleasing emotions simply on account of their appeal to the senses, but rather because the real source of that which is universally recognised as the beautiful is also felt to be THAT WHICH IS THE TRUE; or in other words, the just relation which high art bears to principles of science and the clear perceptions of the intellect.

In general terms, then, it may be said that the most perfect manifestations of the beautiful in art derive their interest from their complete accordance with the laws of the universe, and impress the senses with pleasurable emotions or otherwise in proportion to their harmonious relation with scientific principles.

It is the realization of these truths which in our own time, has been designated by the ill-understood term, " Æsthetics "; and although æsthetics in this sense have been vaguely apprehended by some ancient peoples, and felt to be the underlying principles of the beautiful in art, it is only within our present era of civilization that practical attempts have been made or could become successful, to formulate a science of the beautiful and apply it to various branches of art, more especially to those of colour and tone in relation to painting and music.

The writings of Jean Paul Richter and Hegel— the latter published about the year 1848—are perhaps amongst the best and latest treatises we possess on the science of "Æsthetics," or the beautiful in art; but there are no known works— not excepting those of the above-named authors— which aim to do more than classify effects, rather than search out causes.

In respect to the art of music, considering the paucity of truly analytical works on the subject, it would be difficult to say how or by what precise methods the best thinkers of our own day have arrived at the conviction that the exact principles of mathematics must underlie the most perfect expressions of tone, and that such music as appeals with the most irresistible charm to the cultured ear will invariably be found in harmony with those subtle but immutable laws which ramify in the form of various branches of science throughout the universe.

To those who have been accustomed to think that music was nothing more than ₀a succession of pleasing sounds, and that the cause of the pleasurable sensations they create depends solely upon the taste or fancy of the listener, the above-stated propositions will appear to be a mere meaningless abstraction ; but to the profound student of musical history it will be apparent that the philosophers of all times, when dealing with the subject of music, have invariably realised the necessity of arranging the variety of tones that could be produced, in such order as would bear a true mathematical relation the one to the other.

To arrive at such a result as this, it is evident that many of the most renowned sages of antiquity devoted themselves to patient research and experiment century after century ; and though it is only by aid of modern scientific appliances that the fundamental principles of the beautiful in sound can be fully demonstrated, the search for those principles, and the apprehension of their existence, have occupied the attention of philosophic minds from time immemorial.

On this point Helmholtz, in his elaborate treatise on "Sensations of Tone," says :—

" Melody must express a motion in such a manner that the hearer may clearly and certainly appreciate

the character of that motion by immediate perception. This is only possible when the steps of this
motion, their rapidity and amount, are measurable
by sensible perception. . . . Melodic motion
is change of pitch in time. To measure it perfectly,
the length of time and the distance between the
pitches must be measurable. And this is possible
only for immediate audition on condition that the
alterations both in time and pitch should proceed
by regular and determined degrees. . . . The
alteration in pitch must proceed by intervals,
because motion is not measurable by immediate
perception, unless the amount of space to be
measured is divided off into degrees. . . . We
find the most complete agreement among all nations
that use music at all, from the earliest to the latest
times, as to the separation of certain determinate
degrees of tone, and those degrees form the scale in
which the melody moves. But in selecting the
different degrees of tone (*i.e.,* pitch), deviations of
national taste become apparent. Thus the number
of scales used by different nations at different times
is by no means small."

Our basic proposition being admitted (*i.e.,* that
music is a science, and that the beautiful in sound
depends upon the discovery and application of
scientific laws), it follows that the intervals between
tones of different pitch must conform to scientific
order, and the musician's chief aim must be to search
out those laws, and construct his scales in strict
accordance therewith.

These principles have been perceived by all nations wherein music has been practised, and that long before the era of those Greek philosophers to whom traditional custom so generally ascribes the discovery of tonal laws.

To account for the persistence with which the best writers on the evolution of musical law and order so invariably date back to the periods when Greek philosophy became the authoritative standard of thought in all nations of antiquity, it is necessary to observe that whilst instruments were constructed and music practised in many older countries than Greece, no accurate account of the tonal capacity of those instruments, their methods of tuning, nor the nature of the music performed, can be gathered from the records of any people other than those of the Grecian states.

Music is now acknowledged to have been one of the earliest and most universal expressions of human sentiment and poetic forms of speech; but whilst its origin is lost in the night of antiquity, and the primordial methods of representing musical tones by instrumentation may be truly ranged as amongst "the lost arts," we are compelled to wait for the genesis of such eras of civilisation as incorporate their ideas in the language and limitations of science, before we can successfully trace up the development of tonal laws arranged in the formulæ of art.

It is only amongst the Greeks, who from the remotest periods of their nationality sought to place all their arts on the fundamental basis of scientific law, that we can hope to discover in what order or on what principles the practice of music was developed.

These remarks may serve to explain why we draw our principal illustrations of musical form from Greek methods, theories, and such tabulated records as are handed down to us from classical ages, and why we enter upon the analysis of antique musical science by consulting the vestiges of Greek art, in preference to exploring amongst the almost baseless fabric of musical practices in older nationalities than that of Greece.

Imitative and progressive as Grecian genius undoubtedly was, we have in those elements most probably the conservation of many older and more widespread experiments in our special direction of research, combined with those improved methods which inevitably resulted from the philosophic endeavours of Greek musicians to base musical law and order upon those universal principles observed throughout the order of creation. Still we must premise that, in retracing as far as possible the progress of musical science throughout the most flourishing periods of Greek history, it cannot fail to become apparent that the perfect understanding and application of musical art, and the commencement of its scientific developments, so generally attributed to Greek philosophers, are statements greatly exaggerated, and are far more likely to be due to the proscriptive reverence with which the classical ages of Greek history are regarded than to the actual achievements of Greek musicians in the discovery and systematisation of tonal laws.

We believe we shall be able to show that whilst the best of the Greek philosophers confidently affirmed the necessity for evolving the beautiful in sound from scientific principles, they failed in so many respects in demonstrating their theories that the fundamental laws of melodious progression and harmonious combination were only partially apprehended, and have had to wait for the mechanical appliances of modern times to become resolved into true and beautiful scientific order.

This seems to be a sweeping assertion, and one which the world's conventional deference to classical authority and the custom of attributing to the Greeks the discovery of all the fundamental principles of art and science, may render questionable, nevertheless we believe that a careful study of the succeeding pages will amply prove that many centuries of research and experiment on the part of Grecian philosophers only served to demonstrate that the beautiful in sound was also THE TRUE, *i.e.*, that the principles of musical order must be governed by strict mathematical laws, hence that the beautiful and the true in music are one and inseparable.

To reduce to practice a theory so profound, and a law so apparently immutable, the Greeks continued their experiments from generation to generation, evidently aiming to form such a complete scale of sounds as would be established by the divisions of the tetrachord and octave.

From vague and ill-defined perceptions of possibilities, rather than principles, they were stimulated to a different mode of research by the celebrated Pythagorean question, " Why is consonance determined by the ratios of small whole numbers ? "

Whether this suggestion originated with Pythagoras or merely reflected the opinions of still older Greek sages, we need not now inquire—indeed, the analyses that follow will depict, as far as may be at present possible, the evolution of the various Greek modes and scales.

It is quite probable that Greek mentality—ever bent on discovering one fundamental basis for all

forms of art and science—may have realized that
the number of sounds bounded by the prime and its
octave, must to some extent correspond with the
mathematical unit and its boundary number 9.   Be
this as it may, a certain and assured principle was
perceived in following out the Pythagorean question,
but the application of that principle, as will be
shown, utterly failed in practical demonstration.
Had the Greek musicians actually studied how to
answer the Pythagorean question, and formed their
scale by *ratios of small whole numbers,* such as 1:2,
2:3, 3:4, &c., they would have indeed anticipated
the triumphs of more than twenty centuries later,
and proved themselves entitled to that proscriptive
veneration from the musicians of all subsequent
periods which is now ignorantly rendered to the
prestige of mere classical authority.

By an examination of the ante-Greek periods and
the rudimental scales of other nations, as well as of
the Greek tetrachords and scales, from Orpheus to
Ptolemy, we shall not only trace out the efforts that
the human mind has made to place the art of music
on the fundamental basis of a true science, but we
shall learn, from the failures of the best of philosophers, how inevitably necessary it becomes to construct arts upon the corner stones of science, and
determine that the beautiful must grow out of con-

formity with scientific principles. The analysis of
ancient, and especially of Greek musical scales,
moreover, will teach us another most important
lesson. We shall soon discover that the failure of
the ancients to reduce to practice their theorem that
*consonance depended upon ratios of small whole
numbers between notes,* deprived them of the power
of combining tones into harmony. Their music,
therefore, consisted wholly of such progressions as
we now call melody; their combinations were simply
tones massed together in unison; and all the multi-
form splendours which grow out of instrumental and
vocal combinations, together with the endless range
of harmonies in varieties of accordant sounds, were
unknown. How these have grown, and become
evolved from the five-stringed lyre of the Chinese
and the ante-Pythagorean periods, it will be our aim
to show.

Perhaps one of the grandest triumphs of a
research of this nature will be the final demonstration, that the principle vaguely perceived and
debated upon by the Greeks was in their time,
and still is, the basic truth that underlies the
beautiful in sound, whether in the form of pure
melody or the richest combinations of harmony.
The principles of truth are no less simple than un-
changeable. It may be possible for the intuition of

a savage to perceive them, but it demands all the appliances and resources of civilization to demonstrate and practicalize them. It seems to have required a hiatus of 2,000 years to embody Greek intuition in modern musical art; but if 'in the researches of this last period we have succeeded in establishing the permanency of ideas, the immutability of natural law, and the evolution of the powers of harmony, the labours of past centuries have not been in vain.

Chapter II.

ON THE MOST ANCIENT SERIES OF MUSICAL SOUNDS AND THE METHOD OF EVOLUTION.

IN attempting to trace out the steps by which musical scales, or an orderly succession of musical sounds were originally formulated, it is a matter of equal regret and difficulty to find that no authentic information can be gained concerning the actual tonal characteristics of those instruments devised and used by the most ancient nations, at least by those who are traditionally reported to have practised both vocal and instrumental music anterior to the *historic* times of the Hindoos, Chinese, Assyrians, and Egyptians.

Legendary lore—of which there is a vast abundance amongst Oriental peoples—ascribes the invention of primeval musical instruments now to the accidental discovery of pleasing sounds produced by animal fibres stretched across the shell of a tortoise, now to the same results obtained from striking dried strings formed in the cranial cavities or other remains of animal existences.

The play of the winds on natural æolian harps, the sighing of the breeze through the pipes of reeds and rushes—in a word, the manifold voices of nature —have all been referred to as suggestive hints for the construction of musical stringed and wind instruments, while the accidental resonance of metallic sounds has been considered quite sufficient to account for the formation of those instruments of percussion in which semi-civilized nations appear to take especial delight. But whilst these and many other causes for the invention of musical instruments are alleged, we have to go forward to historical times, and search among ancient writings and monumental remains to discover the names, forms, and something of the characteristics of the earliest musical instruments that were in use.

The Hindoo Vedas—probably the most ancient writings extant—make mention of flutes, lutes, gongs, &c., as forming essential elements in the rites and ceremonials of religious worship. Beyond the names assigned to these instruments, however, we have little or no idea concerning their actual tonal properties.

In describing the early cosmogony of the race, the writers of the Old Testament name Jubal as "the father of all that handle the harp and organ." Here, again, the text conveys no other idea of the instruments alluded to than such as their designation implies; but what the tonal capacity of the harp or organ manipulated by Jubal was, we have no authentic means of determining.

The first definite ideas we can form of the shape and nature of antique musical instruments are derived from the monuments and scriptures of the Chinese, Assyrians, and Egyptians. By these we learn that from very remote periods of time, instruments in the form of a five-stringed lyre were in general use. Besides the above-mentioned nations, we have evidence from undoubted authority that five-stringed instruments resembling the sculptured Assyrian lyre were common among the ancient Malays, the inhabitants of Java and Sumatra, the aborigines of Hudson Bay and New Caledonia, and even amongst the rude Fellah negroes.

Although we have no data to indicate at what remote periods of time these instruments were first constructed, or the methods by which their tones were regulated, we find their use had become universal in the most flourishing periods of Chinese, Assyrian, and Egyptian history.

As there is good reason to believe that changes among the ancients were more infrequent than in modern times, and that each succeeding generation adhered in some measure to the customs of their ancestors, we are justified in supposing that some of the instruments now in use, but the names and origin of which, like the harp, trumpet, cymbal, &c., date back to the remotest antiquity, may furnish us with a clue to the nature and use, if not to the modes of tuning or the tonal capacity, of the most antique musical instruments.

Availing ourselves of these links of connection, we find that there are not only instruments, but also fragments of early Chinese, Scandinavian, and Celtic tunes still extant, which, when reduced to an approximate method of modern notation, and critically examined, present such a number of marked peculiarities as to suggest that they must have been derived from some generally-adopted though at present unknown system.

Helmholtz points out that the result of his examination of a large number of tunes of alleged Pentatonic origin tends to prove that they are for the most part characterized by the deficiency of certain intervals such as compose our diatonic scale ; that "in some tunes of Gaelic origin may be observed the absence of those two sounds which correspond to our third and seventh ; in some others the second and sixth are wanting, while others, again, are deficient in the fourth and seventh, or second and fifth."

The series of sounds discoverable in the ancient music alluded to, it is reasonable to consider, was derived from the tuning of five strings, five pipes of different lengths, or one pipe with air holes bored so as to produce five tones of different pitch, and it is from the universal prevalence of five-toned instruments amongst the most ancient nations that the term pentatonic is derived, and researches for a veritable pentatonic scale have been so often essayed; consequently our first inquiry is, what *tuning* of an instrument would give the observed peculiarities of pentatonic tunes.

Curiously enough we find the answer in the observations and investigations made by Mr. Ellis on the instruments used by a band of Javanese musicians who came to this country in 1882. These were tuned to a Salendro scale, which is a pentatonic, not a heptatonic scale with two notes eliminated, as is the case with some pentatonic scales, but one independent of any heptatonic order.

Mr. Ellis's observations, and his analysis, conducted with great care and experience, are recorded in *vibrations*. He examined three sets of instruments with the following results :—

In the first set		In the second set		In the third set	
1st Note gave	268 vibs.	1st Note gave	272 vibs.	1st Note gave	270 vibs.
2nd	308	2nd	308	2nd	308
3rd	357	3rd	357	3rd	357
4th	411	4th	411	4th	411
5th	470	5th	471	5th	469

C

The discrepancies between the number of vibrations recorded as producing the first and fifth notes in the three sets, are accounted for by the transport of the instruments to a different climate and change of temperature. Mr. Ellis appears to have satisfied himself that the relative pitches were intended to be alike in each set; so we will regard the tuning as intended to conform to about 268, 308, 357, 411, and 469 vibrations.

Now the tuning of these instruments, as the sequel will show, accounts for the peculiarities Helmholtz observes in pentatonic tunes.   In the first place, the intervals are of various widths, and between two only is the width nearly uniform. Between the first and second notes of this scale the ratio is 67 : 77 ; between the second and third the ratio is 44 : 51 ; between the third and fourth the ratio is 119 : 137, and between the fourth and fifth 411 : 469.   Then, if we give the notation C D♯ F G♯ B♭ to the five notes, a tune composed with them would have no interval corresponding to a major second or major sixth.

Analyzing the first evolution of this scale D♯ F G♯ B♭ C, a tune composed with these sounds would be deficient in a major third and a minor seventh.

Analyzing the other scales that could be evolved, it will be found that all the deficiencies remarked by Helmholtz are presented, and as the tuning of the Javanese instruments accounts for the peculiarities noticed in ancient pentatonic tunes, some reliance may be placed at once on the opinion that the *Salendro scale* is the *old Pentatonic scale*.

But there is another remark which must be accounted for in connection with the pentatonic tunes—it is, that a certain number became associated with the bagpipe, and so must have been played, if not on that instrument with its present scale, at all events with its drone accompaniment, which is not likely to have been altered, and which is formed with a sound and its perfect fifth.

To have added the drone below the deepest note of the prime scale to which the Javanese instruments were tuned, is not likely to have been done, because C G below C D♯ F G♯ B♭ would have sounded too distressing to be endured it may be presumed by even Celtic ears, but if the drone be added to the third evolution of the prime scale, viz., to G♯ B♭ C D♯ F, the drone would be G♯ D♯ below G♯ of the scale, and the drone sounds agreeing with G♯ and D♯ of the scale could have been used. That they were, may be inferred from the observation that the old Celtic tunes conform to the scale which is the third evolution of the prime scale, and are deficient, like the scale of the third evolution of the prime, in the intervals of a perfect fourth and seventh.   Hence is further corroborated the opinion that the *Salendro* of the Javanese and the *old Pentatonic* scales are similar.

Regarding it as conclusive that such is the case, we proceed to say that a system which accounts for the tuning of the Salendro scale must necessarily be the *system* of the pentatonic scales.   That system has, however, hitherto escaped the researches of philosophical musicians, and only its requirements have been known.   These determined however, a system can be defined, and if the assumed results meet the requirements of the pentatonic system, the inference is clear that its secret has been disclosed.   In order that there should be no mistake respecting the requirements of the system, we re-iterate and enumerate in full the specialities of the pentatonic scale for which any satisfactory theory of its origin must account.

1. It must explain why the intervening intervals are greater than a major second (204 cents), and less than a minor (316 cents) or even a Pythagorean third (294 cents).

2. Why the widths of the intervening intervals are not uniform.

3. The system must account for an interval being presented rather less than 'a perfect fourth ; for, as

Mr. Ellis observes,* in alluding to the failure of all attempts hitherto made to disclose the system upon which pentatonic instruments were tuned, "No supposition which does not make the fourth flat (that is, *one*-fourth) will represent the result observed."

4. It must account for the lowest note of the series, or one of its evolved series forming a fifth with one of the sounds succeeding the lowest note, on account of certain tunes being associated with the drone of the bagpipe.

5. It must account also in some tunes for the absence of intervals corresponding to our third and sixth, and in other tunes to the deficiency sometimes of the second and fifth, the third and seventh, the second and sixth, and the fourth and seventh.

6. Proceeding on the assumption above stated, that the Javenese Salendro scale of modern times is similar to the pentatonic series of the ancients, a system accounting for the tuning of the latter must exemplify the tuning of the former.

Thus, with a full realization of the requirements of any system that can cover the ground of all the above-stated peculiarities, we submit the following hypotheses, as being the most consistent with the subject under consideration.

---

* *Journal of the Society of Arts*, March, 1885.

Our assumption is, that instead of experimenting with a string to find proportions which would make a graduated series of sounds, the most ancient musicians took a pipe or reed, blowing into which, they could not fail in discovering that the sound most readily produced after the prime was what we call the harmonic seventh, which—if we name the sound of the pipe experimented upon C—we should name B♭.

Thus if the pipe experimented upon was 63 inches, and the sound of the pipe was caused by 268 vibrations, the harmonic seventh would correspond to the pipe shortened to four-sevenths of its length or reduced to 36 inches, and the sound B♭ would be caused by 469 vibrations. Thus, a prime and its harmonic seventh became the extremes of the ancient system, instead of a prime and an octave, as in our modern scale.

This hypothesis, vague as it may appear, is the clue—as what follows will show—to the secret of the pentatonic system, and explains why all attempts have failed to discover that system whilst considering any other sound except the harmonic seventh as one of its extremes.

Our next hypothesis is, that having determined the extremes of the pentatonic system, ancient musicians observed that the interval between the harmonic seventh (the most natural harmonic of the pipe) and that sound next to it in acuteness, viz., the octave to the prime, was similar to that presented when the pipe was shortened by one-half, and when it was shortened to four-sevenths of its length.   This interval is due to the length of pipe which gives the octave being one-eighth shorter than the pipe which gives the sound corresponding to the harmonic seventh, consequently the lengths of the pipes were as 7 to 8.

If the length of the pipe giving the octave were 31½ inches, the length of the pipe whose sound corresponded to the harmonic seventh would be 36 inches.   This interval (ratio 7 : 8) was thus regarded as one proper wherewith to divide the large interval between a prime and its *harmonic seventh*, as was also the interval produced by sub-dividing any length of pipe by two.   Such a division gave first the octave, and then adding to the octave half its length, there was a length three-fourths of the prime.*

The ancient pentatonic musicians had these two intervals wherewith to divide the interval between a prime and its harmonic seventh, viz., one which we shall describe as ratio 7 : 8, and another as ratio 3 : 4.

---

* On page 522 of Helmholtz's "Sensations of Tone," Mr. Ellis expresses the opinion that singing and playing on pipes were probably the first methods of musical practice, so that our idea that pipes, and not strings, were first experimented upon is corroborated.

They then found the note below the harmonic seventh whose ratio was 7 : 8. This they did by lengthening the pipe of 36 inches to 41$\frac{1}{7}$. Then they shortened the prime so as to make another length, with proportion as 3 to 4. This they did by taking three-quarters of the length of the prime. They then had a length of 47$\frac{1}{4}$ inches. Then they increased the length of the pipe of 41$\frac{1}{7}$ inches by adding to it one-third its length, so as to make the proportion as 3 : 4, and had a length 54$\frac{6}{7}$.

The lengths were consequently as 63, 54$\frac{6}{7}$, 47$\frac{1}{4}$, 41$\frac{1}{7}$, and 36 inches, which, giving 268 vibrations to 63 inches, makes a series of 268, 307$\frac{25}{32}$, 357$\frac{1}{3}$, 410$\frac{3}{8}$, 469 vibrations or disregarding unimportant fractions, 268, 308, 357, 411, 469.

Carefully studied, the series thus formulated accounts for the specialities of the pentatonic scale, seeing that it accounts—

1. For the intervening intervals being greater than a major second, and less than a minor third or even a Pythagorean minor third of 294 cents.

2. It accounts for the width of the intervening intervals not being uniform, as the intervening ratios are various.*

---

* If the prime gave 268 vibrations, a major second would give 301$\frac{1}{2}$ vibrations, and a' minor third 321$\frac{3}{4}$. The interval was, it would seem, 308 vibrations, which is greater than a major second by 6$\frac{1}{2}$, and less than a minor third by 13$\frac{3}{4}$ vibrations.

3. It accounts for the observed interval which corresponds very nearly to a perfect fourth, although it is too flat to be so designated; for whereas the difference between two pipes, if a perfect fourth were found, should be 498 cents, the existing difference is 471 cents. This interval will be found between the middle and the last note of the series.

4. The theory above presented also accounts for the existence of a fifth in two of the evolved scales, for, if the series be commenced on F or G♯, and the sounds below it be raised an octave, then F to C would be a perfect fifth, so also would G♯ and D♯.

5. It accounts for those tunes formed from the five sounds constituting the pentatonic scale, being characterized by the absence of sounds corresponding either to our third and sixth, second and fifth, third and seventh, second and sixth, or fourth and seventh. For, in the primal series, which may be noted as C D♯ F G♯ B♭, there are no sounds which correspond to the intervals of a major third and major sixth from C. In the first evolved scale, D♯, F, G♯, B♭, there are no sounds which represent the intervals of a second and fifth from D♯. In the second evolved scale, F, G♯, B♭, C, D♯, the intervals corresponding to a third and seventh are wanting, with F as lowest note. In the third evolved scale, G♯, B♭, C, D♯, F, there are no intervals corresponding to a second and

sixth, reckoning from G♯, and in the fourth evolved scale, B♭, C, D♯, F, G♯, there are no intervals corresponding to a fourth and seventh, with B♭ as lowest note.

6. It accounts for the respective number of vibrations which Mr. Ellis observed to be generated by the five instruments which produced the Salendro scale. That is to say, for the number of observed vibrations being 268, 308, 357, 411, 469, for our figures differ from these so slightly as to be practically identical. In order to compare the results of our hypothetical system with the Javanese pentatonic Salendro scales we give the following diagrams :—

Our Hypothetical Pentatonic Scale.

Notes	First.	Second.	Third.	Fourth.	Fifth
Names	C	D♯	F	G♯	B♭
Vibrations	268	307$\frac{25}{32}$	357$\frac{1}{3}$	410$\frac{3}{4}$	469
Inter. cents	240		258	240	231
Cents from C	0	240	498	738	969

A Salendro Scale according to Mr. Ellis's Observations.

Notes	First.	Second.	Third.	Fourth.	Fifth.
Names	C	D♯	F	G♯	B♭
Vibrations	268	308	357	411	469

Although Mr. Ellis gives 228 as the *mean* number of cents intervening between the first and second notes of the instruments he investigated, he gives 240 cents as a tempered number, and he calls the interval a Pentatone, because it is the fifth part of an octave of 1,200 cents. He also suggests that the pentatonic scale might have been originally formed with the interval he calls a *pentatone*. But if all intervals were pentatones they would be of uniform width, whilst his observations lead to the conclusion that they were not uniform.

Professor Helmholtz suggests that the system of the pentatonic scale might be referred to a mode of reckoning by fifths, but a reckoning by fifths does not account for all the sounds of the pentatonic scale, and Mr. Ellis justly remarks, in his appendix to "Sensations of Tone," that the fifth was an interval only used as a basis of reckoning by later musicians than those who invented the pentatonic scale.

Our Discovered Pentatonic System.

The system which we have above laid out accounts for the special characteristics observed in the pentatonic scales, and considering it most improbable that any two systems differing materially from each other in detail should harmonize so closely in results as to account —the one as well as the other—for the peculiarities of the pentatonic scale, we submit that **the system discovered and detailed above is the method by which the pentatonic scale was originally formed.**

With regard to the special mode of lettering we have adopted, namely, C, D♯, F, G♯, B♭, instead of C, E♭, F, A♭, B♭, we submit that, as the intervening interval—240 cents—between the first and second note is nearer to D as a major second of 204 cents, that to an E♭ of 316 cents the lettering D♯ is preferable to E♭.

And again, the intervening interval of 240 cents between the third and fourth notes is nearer the major second of F—viz., G, 204 cents—than to A♭— 316 cents—so the lettering G♯ is preferable.

In analyzing, however, those Gaelic tunes said by Helmholtz to be deficient in the fourth and seventh, &c., the lettering C, E♭, F, A♭, B♭, should be used, as the Gaelic series, *i.e.*, the third evolution of the foregoing, would then be A♭, B♭, C, E♭, F, the scale according to Helmholtz's lettering without a fourth or seventh, *i.e.*, with the marked peculiarity of Gaelic tunes.

This scale transposed a whole tone lower and played on the black keys of the pianoforte is often quoted as suggestive of the Gaelic pentatonic scale, but as a representation it is very imperfect. The

actual difference between that scale and the five
black notes of the pianoforte tuned to equal tempera-
ment, as all pianos now are, can be seen by looking
at the cents in the following two diagrams :—

*Our Hypothetical Gaelic Pentatonic Scale, like the
third evolution of the Salendro scale, with Helmholtz's mode of lettering, and transposed one whole
tone from previous illustrations :—*

Approximate letters......	$G^{\flat}$	231	$A^{\flat}$	231	$B^{\flat}$	240	$D^{\flat}$	258	$E^{\flat}$	= 960
Intervening cents										
Approximate cents from G♭	0		231		462		702		960	

THE SCALE OF FIVE BLACK KEYS ON THE PIANOFORTE.

Letters	$G^{\flat}$	200	$A^{\flat}$	200	$B^{\flat}$	300	$D^{\flat}$	200	$E^{\flat}$	= 900
Intervening cents										
Cents from G♭.	0		200		400		700		900	

It will be seen that, besides other great differences,
there is a difference of 60 cents between the B♭ and
D♭ of one diagram and the B♭ and D♭ of the other,
so that no proper idea of Gaelic pentatonic tunes
can be formed by playing on the black keys of
a pianoforte.

The peculiarities of all the pentatonic scales, viz.,
those evolved from the prime scale, can be known by
a reference to the cents in the following diagram :—

Intervening cents	C 240 (D♯ E♭) 258 F 240 (G♯ A♭) 231 B♭....Scale without		2nd and 5th.
Cents from C	0 240 498 738 969		
Intervening cents	(D♯ E♭) 258 F 240 (G♯ A♭) 231 B♭ 231 C....	,,	3rd and 7th.
Cents from D	0 258 498 729 960		
Intervening cents	F 240 (G♯ A♭) 231 B♭ 231 C 240 D♯ E♭....	,,	2nd and 6th.
Cents from F	0 240 471 702 942		
Intervening cents	(G♯ A♮) 231 B♭ 231 C 240 D♯ E♭ 258 F....	,,	4th and 7th.
Cents from G♯	0 231 462 702 960		
ntervening cents	B♭ 231 C 240 D♯ E♭ 258 F 240 (G♯ A♮)....	,,	3rd and 6th·
Cents from B♭....	0 231 471 729 960		

If we inquire how far the pentatonic scales formed a series suitable for the ear's delectation untutored to endure badly-related sounds, we find that in the prime scale, there are two simple ratios, that of 3 : 4 between C and F and D♯ and G♯, and that of 4 : 7* between C and B♭

In the first evolved scale there is the ratio of 3 : 4 between D♯ and G♯, and the ratio of 2 : 3 between F and C.

In the second evolved scale there is the ratio of 2 : 3 between F and C, and the ratio of 2 : 3 between G♯ and D♯.

In the third evolved scale there is the ratio of 2 : 3 between G♯ and D♯ and the ratio of 3 : 4 between C and F.

In the fourth evolved scale there is the ratio of 3 : 4 between C and F and between D♯ and G♯.

So that consonances can be formed in the pentatonic scales, and simple ratios are to a certain extent exemplified, whilst Celtic musical ears are to a certain extent vindicated.

---

* Note.—The sounds with ratios 4 : 7 are natural notes upon horns and trumpets, but this ratio, though simpler than some, is not introduced into our diatonic scale, nor is it usually cited as a simple ratio.

CHAPTER III.

THE ANCIENT GREEK SERIES OF SOUNDS.

HAVING considered in some detail the nature and probable origin of that succession of sounds formed of five tones in general use amongst ancient nations prior to the Greek period, and distinguished as the "pentatonic scale," we shall now proceed to treat of Greek music in particular, and endeavour to analyse those scales or successions of sounds which are supposed to have been in use amongst the Greeks at different epochs of their nationality.

In the absence of direct and authentic information concerning the earliest known successions of sounds adopted by Greek musicians, we can only surmise what they may have been from considering the capabilities of the instruments in use.

The first of these to which both history and tradition refer was the *Three-stringed Lyre*, said to have been constructed by Amphion, the reputed "Son of Jupiter and Antiope," but certainly one of the earliest Greek musicians who emerges from the mists of antiquities into the light of history. Assuming that the three - stringed lyre must

have been tuned in the simplest manner, it seems
probable that the second and third strings were
respectively three-fourths and one-half of the length
of the first.

If we call the first or prime tone A we have the
following succession A D a (ratios from A 3:4, 1:2).
These intervals, simple as they are, doubtless represent the earliest varieties of tones common amongst
older nations than the Greeks, whose plastic mentality induced them to imitate, at the same time
that they improved upon the methods of other
peoples.

Following upon the three-stringed lyre, the
most ancient of Greek stringed instruments appears
to have been that of the celebrated Orpheus, whose
period is generally placed at about 1200 B.C., and
whose achievements as a musician occupy a distinguished place in classical history.

* Boethius says, on the authority of Nichomachus,
that the Orphic Lyre consisted of four strings, tuned
to the extremes of two disjunct tetrachords. These
would have produced sounds represented by the
following succession :—

Intervening ratios	A 3:4	D 8:9	E 3:4	a
Intervening cents	498	204	498	
Ratios from A	1	3:4	2:3	1:2
Cents from A	0	498	702	1200

* De Musica Lib. I., cap. 20.

The term "tetrachord" signifies the interval of the perfect fourth, but the nomenclature is often used indiscriminately to designate the interval as well as the succession of sounds within the interval.

Rousseau, an assured authority on the subject, says, "The number of strings from which the tetrachord derived its name is not essential; the title may be applied to the interval itself." As regards the tetrachordal system, if it did not grow into use from the construction of these and four-stringed instruments, the universal adoption of the interval by early Greek musicians corresponds with the date of those instruments.

All classical history tends to show that even in the pre-Pythagorean periods of Greek art, instruments with three, four, and even eight strings were known, but as we have no reliable information concerning the succession of sounds produced, we must still resort to reasonable hypotheses as to the method of tuning these instruments.

Recurring again to the Orphic lyre, the sounds of which we illustrate thus :—

Intervening ratios ...	A 3 : 4	D 8 : 9	E 3 : 4	A
Intervening cents ...	498	204	498	
Ratios from A.........	1	3 : 4	2 : 3	1 : 2
Cents from A	0	498	702	1200

it will be observed that the second and third notes are divided by what we now call a major second,

D

ratio 8 : 9, and this interval appears from the earliest times to have been adopted by the Greeks as a natural mode of tonal progression, just as the ratio 7 : 8 appears to have been adopted by the pentatonic musicians.

Taking the major second as an interval wherewith to make other divisions of the tetrachord, it may be assumed, that when a tone was desired below, say—D, the string was tuned to the same distance of pitch as that occurring between D and E. This would give C, and the succession would appear as follows, and represent *the five-stringed lyre :*—

	1st Tetrachord.			2nd Tetrachord.	
	A	C	D	E	a
Intervening ratios ...	27 : 32	8 : 9	8 : 9	3 : 4	
Intervening cents ...	294	204	204	498	
Ratios from D to A...	3 : 4	8 : 9	1	—	—
Ratios from a to E ...	—	—	—	3 : 4	1
Ratios from A.........	1	27 : 32	3 : 4	2 : 3	1 : 2
Cents from A	0	294	498	702	1200

Continuing the plan of reckoning by a major second below a note already established, and reckoning a major second below C, the note B♭ would be presented. This gave the following succession which we may reasonably suppose represented

THE SIX-STRINGED LYRE.

	1st Tetrachord.				2nd Tetrachord.	
	A	B♭	C	D	E	a
Intervening ratios ...	243 : 256	8 : 9	8 : 9	8 : 9	3 : 4	
Intervening cents ...	90	204	204	204	498	
Ratios from D to A..	3 : 4	64 : 81	8 : 9	1	—	—
Ratios from a to E..	—	—	—	—	3 : 4	1
Ratios from A	1	243 : 256	27 : 32	3 : 4	2 : 3	1 : 2
Cents from A	0	90	294	498	702	1200

THE TERPANDER LYRE.

Terpander, a musician, of scarcely less repute than Orpheus, and whose artistic exploits date back to 600 years antecedent to our era, is credited by Plutarch and Athenas with having been the originator of the seven-stringed lyre, an improvement upon still earlier instruments of quite sufficient importance to mark an epoch in the history of musical art and account for the high distinction associated with the name of Terpander as a musician.

To arrive at the nature of those sounds produced by the seven-stringed lyre, we have only to apply the same principle as that indicated before, and by obtaining a note at the distance of the major second below the octave we have the sound G completing the series of seven notes.

THE TERPANDER SERIES OF SEVEN NOTES.

	A		B♭	C	D	E	G	a
Intervening ratios..		243 : 256	8 : 9	8 : 9	8 : 9	27 : 32	8 : 9	
Intervening cents ..		90	204	204	204	294	204	—
Ratios from D to A	3 : 4		5 : 6	8 : 9	1	—	—	—
Ratios from a to E.	—		—	—	—	3 : 4	8 : 9	1
Ratios from A......	1		243 : 256	27 : 32	3 : 4	2 : 3	9 : 16	1 : 2
Cents from A.......	0		90	294	498	702	990	1200

In support of the above hypothesis respecting the Terpander scale, we have the opinion of Professor Helmholtz who says ("Sensations of Tone," page 257):

"Terpander, who played a conspicuous part in the development of Greek music, added a seventh string to the former cithera of six strings, and a scale composed of a tetrachord and a trichord, having the compass of the octave, and tuned thus : E, F, G, A, B—D, E.   This, when transposed a fifth lower, agrees with the series given as A, B♭, C, D, E—G, A."

It was by the aid of the seven-stringed lyre, termed a " heptachord," that Terpander won the crown of victory in the Carnean musical contest of 676 B.C., besides gaining four prizes in the Pythian games and achieving other feats of the highest musical celebrity for the time in which he lived.

Referring to Terpander's invention of the cithera and the heptachord, Strabo quotes a couplet supposed to be an extract from the numerous works of the great musician himself, of which the following lines are an English version :—

> " The tetrachord restraints we now despise,
> The seven-stringed lyre a nobler strain supplies."*

THE EIGHT-STRINGED LYRE.

In process of time we have evidence that an eighth string was added to the Terpander lyre. According to Boethius this notable improvement was effected by Lychon of Samos, though Nichomachus attributes it to Pythagoras.   Be this as it may,

---

* Plutarch says that Terpander set his "nomes" in Hexameter verse to music, and Clemens Alexandrine alleges that he even "set the laws of Lycurgus to music."

the question still remains as to what note it might be which was added to the Terpander series. It could not have been the octave, if the scale already given (which includes the octave) be correct.

Nichomachus * in quoting Philolaus, says that "From the hypate E to mese A was a fourth. From mese A to nete E was a fifth. From the nete E to trite B was a fourth, and from trite B to hypate E was a fifth." So, evidently, the octave to the prime was not the sound for which an additional string was required.

In reviewing this series it seems more probable that Lychon formulated this scale than Pythagoras.

Pythagoras would doubtless have contributed a perfect fifth or a major seventh, for these special intervals accord with his recognised ideas concerning the construction of a scale. But the Terpander series comprised a perfect fifth and a minor seventh, so the addition of a fifth and seventh to the scale was not required.

These considerations lead to the belief that Nichomachus desired to imply that Pythagoras formed his scale of eight notes, say, C, D, E, F, G, A, B, in contradistinction to the Terpander scale which consisted of seven notes only, say, C, D♭, E♭, F, G, B♭, C.

---

* Meibomius, page 17.

If Lychon of Samos added another string to the Terpander cithera, he in all probability arranged the second tetrachord like the first, namely, established a note (ratio 8 : 9), say A♭ below the second note, reckoning from right to left of the upper tetrachord. This would make the Terpander scale C, D♭, E♭, F, G, A♭, B♭, instead of C, D♭, E♭, F, G, B♭, C. As the weight of testimony points to Lychon of Samos as the originator of the first eight stringed cithera, it is not necessary to follow the too popular and often erroneous custom of attributing this—together with all manner of other improvements in science or philosophy—to the renowned Samian Sage, Pythagoras. Believing that the note added to the Terpander lyre of seven strings was the sixth note of the scale we give as follows :—

THE TERPANDER LYRE OF EIGHT STRINGS.

	First Tetrachord.				Second Tetrachord.			
	A	B	C	D	E	F	G	a
Intervening ratios..	243 : 256	28 : 9	8 : 9	8 : 9	243 : 256	8 : 9	8 : 9	
Intervening cents..	90	204	204	204	90	204	204	
Ratio from D to A..	3 : 4	64 : 81	8.9	1				
Ratio from a to E..	—	—	—	—	3 : 4	64 : 81	8 : 9	1
Ratios from A......	1	243 : 256	27 : 32	3 : 4	2 : 3	81 : 128	9 : 16	1 : 2
Cents from A......	0	90	294	498	702	792	996	1200

THEORIES CONCERNING GREEK MUSICAL ART.

From the frequent mention of Greek musical instruments in classic poems and history, especially in connection with the seven-stringed Terpander lyre, it may be inferred that the same method by which the interval of a perfect fourth or the tetrachord was divided was followed more or less completely in filling up the second tetrachord.

No information respecting the plan adopted, however, has been handed down to us, and nothing is decisively known concerning the divisions of the tetrachord even in regard to the Terpander scale, although it is supposed to exemplify the methods of division observed in the Dorian tetrachord, which has been preserved, and the order of which will presently be given.

Some of the characteristics of other tetrachords besides the Dorian can also be examined. Thus, though we cannot authoritatively determine what the Greek systems were, we may draw satisfactory inferences concerning them from a careful study of the tetrachords and their sub-divisions.

THE TETRACHORDAL SYSTEMS.

We commence our analysis of the tetrachordal systems with

THE DORIAN TETRACHORD.

	A 243 : 256	B♭ 8 : 9	C 8 : 9	D
Intervening ratios...				
Intervening cents ...	90	204	204	
Ratios from D	3 : 4	4 : 5	8 : 9	1
Ratios from A	1	243 : 256	27 : 32	3 : 4
Cents from A	0	90	294	498

The Dorian, or Doric, is considered to be the most ancient of the tetrachordal divisions, of which a great many were formulated at different periods of time, and by different musicians. It is true that only a limited number were in practical use, *i.e.*, only a small proportion were applied to the tuning of instruments, the theoretical sub-divisions, or as the Greeks termed them "ἄλογα" (not having a ratio expressible by whole numbers), only serving to exemplify the many ways by which the tetrachord could be divided.

The six principal tetrachordal divisions [*] are supposed to have been—

> THE DORIAN,
> THE OLYMPOS,
> THE DIATONIC,
> THE LYDIAN,
> THE PHRYGIAN,
> THE CHROMATIC,
> THE ENHARMONIC.

We now proceed to consider what successions of tones were represented by these tetrachords, and how they were evolved.

---

[*] In his Table of old Greek tetrachords, page 20, of Ellis's Appendix to Helmholtz's "Sensations of Tone," the writer includes the Didymus tetrachord in his list. As Suidas makes Didymus contemporary with Nero, the scale he formulated is not here classed with the pre-Pythagorean Greek tetrachords.

The Dorian tetrachord has already been illustrated. The Olympos derived its name from its originator, who was reputed some 620 years B.C. to have introduced from Asia a five-toned flute, the notes of which are not known. It is to be presumed, however, that these consisted of whole or wider tones, as Olympos, with a view to propitiate the well-known Greek taste for chromatic intervals, retuned his flute in order to produce semitones.

This item of history renders it probable that the original Olympos flute series represented the notes of the ancient pentatonic scale (*vide* "Pentatonic Scales," chapter II.)

To effect the desired change in the notes of his flute and introduce the desideratum of semitones, it is evident Olympos did not adopt the Doric series, for we are informed that he changed or "improved" upon it. To determine the mode by which he arrived at the succession he finally adopted, our hypothesis is that he sub-divided the Grecian tetrachord by using the ratio 4 : 5 and adopting the defect — the semitone, ratio 15 : 16, a method formulated two centuries later by Archytas. The semitone evolved as above indicated was of a far more pleasing kind than that furnished by the Doric ratio 243 : 256 (cents 90), and yet it established such a series as would accord with the Greek taste for semitones.

According to our modern notation, the letters which would represent the above plan would be

A   B♭   D

To obtain five notes Olympos appears to have set up a conjoint second tetrachord on D and divided it like the first, thus :—

THE OLYMPOS SERIES.

Conjoint Tetrachords.

	A	B♭	D	E♭	G
Intervening ratios ...	15 : 16	4 : 5	15 : 16	4 : 5	
Intervening cents ...	112	386	112	386	
Ratios from D to A...	3 : 4	4 : 5	1	—	—
Ratios from G to D...	—	—	3 : 4	4 : 5	—
Ratios from A.........	1	15 : 16	3 : 4	45 : 64	9 : 16
Cents from A	0	112	498	610	996

Some musical historians allege that Olympos changed the Doric tetrachordal series into one of five notes, but in deference to the then established custom of dividing tetrachords it is more plausible to assume the above series to have been the correct one.

We thus see how, before the time of Archytas, the interval with the ratio 15 : 16 was brought into notice, viz., as the defect of the major third. Helmholtz says Archytas was the first to reckon this interval of the major third from the upper note of the tetrachord and give its ratio as 4 : 5, but Archytas, who lived about two centuries later than Olympos, probably aimed at formulating the idea which had already been applied, instead of originating the method in question.

In this, as in many other instances in classic history, discoveries in art and science are very frequently associated more directly with the individual capable of systematizing ideas into form and order than with the originator, and hence the many obstacles which present themselves to obstruct the path of crucial research and investigation.

The establishment of the Olympos tetrachord with the intervening notes above referred to, introducing the ratio 15 : 16, gave rise, it is presumed, to the *Diatonic tetrachord*, for this tetrachord is a completion of Olympos's by the addition of a note (ratio 8 : 9), below the highest note, as shown by Archytas, Eratosthenes, and other Greek musicians. The consequence is, no further notice is taken of Olympos's tetrachord, which became merged into

THE DIATONIC TETRACHORD.

	A	Bb	C	D
Intervening ratios ...	15 : 16	9 : 10	8 : 9	
Intervening cents ...	112	182	204	
Ratios from D to A...	3 : 4	4 : 5	8 : 9	1
Ratios from A to D...	1	15 : 16	27 : 32	3 : 4
Cents from A	0	112	294	498

The *diatonic tetrachord* appears in turn to have given rise to two other principal tetrachords by simply inverting each successive set of ratios.

Thus, the series of ratios for the diatonic, from grave to acute, being 15 : 16, 9 : 10, 8 : 9. The first inversion was 9 : 10, 8 : 9, 15 : 16. The second inversion was 8 : 9, 15 : 16, 9 : 10.

The first inversion of the series gives

THE LYDIAN TETRACHORD.

which we set up with ratios 9 : 10, 8 : 9, 15 : 16.

Intervening ratios ...	A 9 : 10	B 8 : 9	C♯ 15 : 16	D
Intervening cents ...	182	204	112	
Ratios from D.........	3 : 4	5 : 6	15 : 16	1
Ratios from A	1	9 : 10	4 : 5	3 : 4
Cents from A	0	182	386	498

The second inversion of the series gives

THE PHRYGIAN TETRACHORD.

Intervening ratios ...	A 8 : 9	B 15 : 16	C 9 : 10	D
Intervening cents ...	204	112	182	
Ratios from D.........	3 : 4	5 : 6	8 : 9	1
Ratios from A.........	1	8 : 9	5 : 6	3 : 4
Cents from A	0	204	316	498

THE OLD GREEK CHROMATIC TETRACHORD.

The diatonic tetrachord may have also given rise to the *old Greek Chromatic Tetrachord* for the ratios 3 : 4, 4 : 5, (read from right to left) are adhered to by Didymus in all his tetrachords. It is true Didymus was a comparatively modern Greek, but he favoured the simple ratios of the old Greeks, and disdained to be influenced by the Pythagoreans.

Didymus makes the chromatic by lowering the diatone (C) in the diatonic tetrachord 112 cents, and changing the ratio 8:9 to 5:6 ; thus according to Didymus the old Greek chromatic genera scale should be formed thus—

DIAGRAM OF THE OLD GREEK CHROMATIC TETRACHORD.

	A	B	C	D
Intervening ratios...	15 : 16	♭24 : 25	♭5 : 6	
Intervening cents...	112	70	316	
Ratios from D	3 : 4	4 : 5	5 : 6	1
Ratios from A	1	15 : 16	9 : 10	3 : 4
Cents from A	0	112	182	493

THE ENHARMONIC TETRACHORD.

The diatonic tetrachord it is also presumed gave rise to the *enharmonic tetrachord,* for the ratios 4 : 5 and 15 : 16 of the diatonic are preserved in it according to Archytas and Didymus tetrachords. To obtain the enharmonic interval, the diatone C was lowered, but what was its ratio cannot be stated, for under the Pythagorean influence the old Grecian genera scales fell into disuetude, and the ratio for the Enharmonic interval was lost. Didymus divided the ratio 15 : 16 as nearly equally as possible, and thus gave the enharmonic interval the ratio 24 : 31 from the highest note. But the normal ratio for the enharmonic interval was as unknown to Didymus as to modern musicians. An attempt to recover this lost division is thus alluded to in "Photii Bibliotheca" (edit. Berthetia Rothomay):

"Asclepiodurus shifted and transposed frets to the number of not less than 220, but failed to find the enharmonic, though he divided and executed the other two genera."

If an hypothesis on the subject might be allowed, the method of producing the enharmonic interval, possibly was, to fill in the interval between A and B♭ with a note whose interval ratio to B♭ was the same as the interval ratio between B♭ and C♭, viz., 24 : 25, cents 70.

This would make the enharmonic interval ratio reckoned from the highest note, 96 : 125.

The *Hypothetical Enharmonic Tetrachord* would then be*

	A	C♭♭	B♭	D
Intervening ratios...	125 : 128	24 : 25	4 : 5	
Intervening cents...	42	70	386	
Ratios from D	3 : 4	96 : 125	4 : 5	1
Ratios from A	1	125 : 128	15 : 16	3 : 4
Cents from	0	42	112	498

The enharmonic tetrachord of Didymus is, however as follows :—

	A	C♭♭	B♭	D
Intervening ratios...	31 : 32	30 : 31	4 : 5	
Intervening cents...	55	57	386	
Ratios from D	3 : 4	24 : 31	4 : 5	1
Ratios from A	1	31 : 32	15 : 16	3 : 4
Cents from A	0	55	112	498

* As the Chromatic and Enharmonic Genera scales appear to have been formed by lowering the *Diatone* C of the Diatonic scale, we have preserved the letter C in the Enharmonic scales. C double flat is of course lower in pitch than B♭.

Whilst Archytas' enharmonic tetrachord is

Intervening ratios... Intervening cents...	A 27:28 63 3 : 4 1 0	C♭♭ 35:36 49 7 : 9 27 : 28 63	B♭ 4:5 386 4 : 5 15 : 16 112	D 1 3 : 4 498

From our analysis of the above-mentioned tetra-chords, it will be seen that the titled tetrachords, namely, the Phrygian and Lydian, as also the chromatic and enharmonic tetrachords had *one common origin*, namely, the Olympos tetrachord merged into the Diatonic. This *discovery*, we think, will be found not only interesting but of remarkable value, for it removes all doubt as to the proper divisions for the Phrygian and Lydian tetrachords.

Notwithstanding the absence of definite proof concerning the methods employed to obtain the intervening notes in the tetrachords, there is no doubt that some general principles were acted upon, and one of these all competent authorities concur in defining as a rule by which the product of the divisions must necessarily equal the perfect fourth. This theorem will be found illustrated in all tetrachords the intervening intervals of which are equal to the ratio 3 : 4, viz., that of the perfect fourth.*

---

* The extremes of the old Greek never varied. The product of the ratios of the Lydian tetrachord reduced to 3 : 4 is here shown, as an example how the product of all tetrachords is acquired and reduced $\frac{9}{10} \times \frac{8}{9} \times \frac{15}{16} = \frac{1080}{1440} = \frac{3}{4}$.

THE OLD GREEK SCALE EXTENDED FROM FOUR NOTES TO SEVEN.

The constructors of the Greek tetrachords, imitating the design apparent in the completed Terpander lyre, extended the tetrachordal series from four to seven notes within the octave. These were established by founding a second tetrachord on the upper note of the first. This system is that of conjoint tetrachords, in contradistinction to the system of disjunct tetrachords, in which the second tetrachord commences on the note a major second beyond the upper note of the first tetrachord, *i. e.*, on the perfect fifth to the prime. The following are examples of both methods—

	First Tetrachord.	Second Tetrachord.
Conjoint tetrachords	A–D	D–G
Disjoint tetrachords	A–D	E–A

After establishing the desired succession of seven notes within the octave, by setting up a conjoint tetrachord, the Greek musicians, following out the same method, set up a tetrachord on the octave to the lowest note of the series, also on the fourth of the octave, and by continuing this method extended the succession at pleasure. The following illustrations show the extension to seven notes within the octave by the use of conjoint tetrachords, which was the *old Grecian plan*, whilst it will be seen that the use of disjunct tetrachords was the later and Pythagorean plan.

THE DORIC SCALE OF SEVEN NOTES.

	First Tetrachord			Second Tetrachord			
	A	B	C	D	E	F	G
Intervening ratios	243 : 256	b8 : 9	8 : 9	243 : 256	b8 : 9	8 : 9	
Intervening cents	90	204	204	90	204	204	
Ratios from D ...	3 : 4	64 : 81	8 : 9	1	—	—	—
Ratios from G ...	—	—	—	3 : 4	64 : 81	8 : 9	1
Ratios from A ...	1	243 : 256	27 : 32	3 : 4	729 : 1024	81 : 128	9 : 16
Cents from A ...	0	90	294	498	588	792	996

THE PHRYGIAN.

	First Tetrachord			Second Tetrachord			
	A	B	C	D	E	F	G
Intervening ratios ...	8 : 9	15 : 16	9 : 10	8 : 9	15 : 16	9 : 10	
Intervening cents ...	204	112	182	204	112	182	
Ratios from D	3 : 4	27 : 32	9 : 10	1	—	—	—
Ratios from G	—	—	—	3 : 4	27 : 32	9 : 10	1
Ratios from A	1	8 : 9	5 : 6	3 : 4	2 : 3	5 : 8	9 : 16
Cents from A	0	204	316	498	702	814	996

THE LYDIAN.

	First Tetrachord			Second Tetrachord			
	A	B	C	D	E	F	G
Intervening ratios..	9 : 10	8 : 9	#15 : 16	9 : 10	8 : 9	#15 : 16	
Intervening cents ..	182	204	112	182	204	112	
Ratios from D.. ...	3 : 4	5 : 6	15 : 16	1	—	—	—
Ratios from G......	—	—	—	3 : 4	5 : 6	15 : 16	1
Ratios from A......	1	9 : 10	4 : 5	3 : 4	27 : 40	3 : 5	9 : 16
Cents from A	0	182	386	498	680	884	996

Mr. Ellis, in his illustrations of the most ancient Greek scales extended to seven sounds and an octave, makes the Greeks adopt disjunct tetrachords, thus giving to the Lydian scale a perfect fifth and major seventh. Without contending that the perfect fifth

E

was an interval peculiar to the Pythagorean and later systems, all authorities agree in ranking the major seventh as such. That the minor seventh is the characteristic of the old Greek scales is universally admitted. The plan of extending the old Greek scales with disjunct tetrachords presents a major seventh, and this is in itself fatal to the assumption that *disjunct* rather than *conjunct* tetrachords were used in the succession of the old Greek seven-toned scale within the octave.

By way of further illustrating the point above raised, we herewith give two diagrams of the Lydian tetrachord, one with the conjunct, the other with the disjunct method :—

LYDIAN WITH CONJUNCT TETRACHORDS.

Two Conjunct Tetrachords.

	A	B	C#	D	E	F#	G	with a added
Intervening ratios.. / Intervening cents ..	9:10 182	8:9 204	15:16 112	9:10 182	8:9 204	15:16 112	204	
Cents from A	0	182	386	498	680	884	996	1200

LYDIAN WITH DISJUNCT TETRACHORDS.

Two Disjunct Tetrachords.*

	A	B	C#	D	E	F#	G	a
Intervening ratios.. / Intervening cents ..	9:10 182	8:9 204	15:16 112	8:9 204	9:10 182	8:9 204	15:16 112	
Cents from A	0	182	386	498	702	884	1088	1200

* The above Lydian, with disjunct tetrachords and G# 1088 cents, is neither the old Grecian nor the later Pythagorean Lydian, but approximates to the modern major scales.

OF THE MODES.

WE must now consider those particular tetrachordal successions of seven notes within the octave, which were called " Modes." These modes each commenced on a distinct and separate initial note—*i.e.*, a note of a different pitch from each other, selected from one primal mode.

According to Pliny, in his treatise on Greek music, the first in the order of the modes was the *Phrygian*, *i.e.*, the scale formed by conjunct *Phrygian* tetrachords. The second mode was the old *Dorian* set up on the second note of the Phrygian mode. The third was the *Lydian* set up on the third note of the Phrygian mode. Each of these modes had an attendant mode formed on the fourth note above their initial notes, and distinguished from their original or parent modes by the prefix "hypo." Thus the Phrygian, Dorian, and Lydian, duplicated by their attendant hypos of the same names,* formed six modes, to which a seventh was added, formed on the seventh note of the Phrygian mode, and with the Lydian succession, hence named Mixolydian.

---

*It may seem strange that these scales, which were more acute in pitch than the others, should have been named with the prefix hypo (*below*). The term, however, was deemed applicable, because on stringed instruments the more acute sounds were produced by moving the hand down the neck of the instrument towards the bridge.

If we adopt A as the initial note of the Phrygian mode, we have the following succession :—

PHRYGIAN MODE.

	A	B	C	D	E	F	G
Intervening ratios ...	8:9	15:16	9:10	8:9	15:16	9:10	
Intervening cents ...	204	112	182	204	112	182	
Ratios from D to A...	3:4	5:6	9:10	1	—	—	—
Ratios from G to D...	—	—	—	3:4	5:6	9:10	1
Ratios from A.........	1	8:9	5:6	3:4	2:3	5:8	9:16
Cents from A	0	204	316	498	702	814	996

DORIC MODE.

	B	C	D	E	F	G	A
Intervening ratios....	243:256	8:9	8:9	243:256	8:9	8:9	
Intervening cents	90	204	204	90	204	204	
Ratios from E to B ..	3:4	64:81	8:9	1	—	—	—
Ratios from A to E ..	—	—	—	3:4	64:81	8:9	1
Ratios from B	1	243:256	27:32	3:4	729:1024	81:128	9:16
Cents from B	0	90	294	498	588	792	996

LYDIAN MODE.

	C	D	E	F	G	A	B♭
Intervening ratios .	9:10	8:9	15:16	9:10	8:9	15:16	
Intervening cents ..	182	204	112	182	204	112	
Ratios from F to C	3:4	5:6	15:16	1	—	—	—
Ratios from B♭ to F	—	—	—	3:4	5:6	15:16	1
Ratios from C......	1	9:10	4:5	3:4	27:40	3:5	9:16
Cents from C	0	182	386	498	680	884	996

Treating the "Hypo" scales, as having been founded by placing their first or initial note on the fourth of their originals, we have the four following additional modes :—

HYPO-PHRYGIAN MODE.

	D	E	F	G	A	B♭	C
Intervening ratios.	8:9	15:16	9:10	8:9	15:16	♭9:10	
Intervening cents.	204	112	182	204	112	182	
Ratios from G to D	3:4	5:6	9:10	1	—	—	—
Ratios from C to G	—	—	—	3:4	5:6	9:10	1
Ratios from D ...	1	8:9	5:6	3:4	2:3	5:8	9:16
Cents from D......	0	204	316	498	702	814	996

THE HYPO-DORIAN MODE.

	E	F	G	A	B♭	C	D
Intervening ratios	243:256	8:9	8:9	243:256	8:9	8:9	
Intervening cents	90	204	204	90	204	204	
Ratios from A to E	3:4	64:81	8:9	1			
Ratios from D to A	—	—	—	3:4	64:81	8:9	1
Ratios from E ...	1	243:256	27:32	3:4	729:1024	81:128	9:1
Cents from E......	0	90	294	498	588	792	996

THE HYPO-LYDIAN MODE.

	F	G	A	B♭	C	D	E♭
Intervening ratios ..	9:10	8:9	15:16	9:10	8:9	15:16	
Intervening cents	182	204	112	182	204	112	
Ratios from B♭ to F	3:4	5:6	15:16	1			
Ratios from E♭ to B♭	—	—	—	3:4	5:6	15:16	1
Ratios from F	1	9:10	4:5	3:4	27:40	3:5	9:16
Cents from F	0	182	386	498	680	884	9966

THE MIXO-LYDIAN MODE.

	G	A	B	C	D	E	F
Intervening ratios.	9:10	8:9	15:16	9:10	8:9	15:16	
Intervening cents.	182	204	112	182	204	112	
Ratios from C to G	3:4	5:6	15:16	1			
Ratios from F to C	—	—	—	3:4	5:6	15:16	1
Ratios from G ...	1	9:10	4:5	3:4	27:40	3:5	9:16
Cents from G......	0	182	386	498	680	884	996

Thus the interval of an octave was filled up with the initial sounds of the seven modes. It must be added that in later Greek periods the term "Hypo" was used in a different sense, and applied to designate the scale of a fourth below, whilst the term "Hyper" was introduced to signify a scale a fourth above the parent scale.

Ptolemy appears to have sanctioned the later application of the word "Hypo" by adopting it in his "Harmonicorum" to indicate a scale a fourth below the parent scale. The ecclesiastical musicians also followed this plan, naming their fourth below a titled scale " Hypo." Thus the ecclesiastical Dorian commenced on D, and gave rise to the appellation of the ecclesiastical Hypo-Dorian founded on A, a fourth below. Alypius, in giving diagrams of all the genera scales, adopts a like order in applying these terms.

THE GENERA SCALES.

Amongst Greek musicians the diatonic, chromatic, and enharmonic scales were distinguished by the term *genera*. Thus there were the diatonic-genera, chromatic-genera, and enharmonic-genera scales.

In order to set up the genera scales of seven notes, conjoint tetrachords were used as with the titled scales. Thus the *diatonic genera scale* formed with two conjoint tetrachords was :—

	1st Tetrachord.			2nd Tetrachord.			
	A	B	C	D	E	F	G
Intervening ratios	15 : 16	b9 : 10	8 : 9	15 : 16	b9 : 10	8 : 9	
Intervening cents	112	182	204	112	182	204	
Ratios from D ...	3 : 4	4 : 5	8 : 9	1	—	—	—
Ratios from A ...	1	15 : 16	27 : 32	3 : 4	45 : 64	81 : 128	9 : 16
Cents from A ...	0	112	294	498	610	792	996

The *chromatic genera scale* formed with two conjoint tetrachords, would be according to Didymus's tetrachord :—

	1st Tetrachord.			2nd Tetrachord.			
	A	**B**♭	**C**♭	**D**	**E**♭	**F**♭	**G**
Interv. ratios..	15:16	24:25	5:6	15:16	24:25	15:16	
Interv. cents...	112	70	316	112	70	316	
Ratios from D.	3:4	4:5	5:6	1	—	—	—
Ratios from A.	1	15:16	9:10	3:4	45:64	27:40	9:16
Cents from A.	0	112	182	498	610	680	996

The *enharmonic genera scale* formed with two conjoint enharmonic tetrachords would be according to Didymus's tetrachord :—

	1st Tetrachord.			2nd Tetrachord.			
	A	**C**♭♭	**B**♭	**D**	**F**♭♭	**E**♭	**G**
Interv. ratios..	31:32	30:31	4:5	31:32	30:31	4:5	
Interv. cents....	55	57	386	55	57	386	
Ratios from D..	3:4	24:31	4:5	1	—	—	—
Ratios from A...	1	31:32	15:16	3:4	93:128	45:64	9:16
Cents from A...	0	55	112	498	553	610	996

The genera scales could be set up on the lowest note of a system, say on A, as in our illustration, also upon the octave, and upon any note of the diatonic scale. We state this on the authority of Gaudentius, who in his " *Philosophii Harmon. Introd.*," says :—

." Sometimes we take some one note between the lowest note (*Proslambanemos*) and its octave for the beginning of the system (*i.e.*, for the formation of the genera scale on a higher note), and to this we tune (*i.e.*, set up the diatonic genera scale). But it is necessary that in every system (*i.e.*, in every move) the same relation which exists between the octave in one system and the octave in the other, or between the keynote in one system and the keynote in another, should exist between the notes of the same name in the two systems respectively, and between the entire of one system and the entire of the other."

It is evident from the foregoing that the plan adopted was identical with that which we call the movable Do system.

It really was the *System Enharmonic*, which must not be confused with the term enharmonic tetrachord, or enharmonic scales, though the system comprised such tetrachords and scales. The *System Enharmonic* was that which insisted upon all scales being tuned in the manner alluded to by Gaudentius.

Colonel Perronet Thompson, in "Just Intonation," says,—

"The essence of what the ancients called the *enharmonic* consisted in transferring the divisions as determined for a single key (or scale) to a variety of keys (that is, notes) by beginning again on some of the previous divisions."

Quintillian (lib. iii.) says,—

"The *enharmonic*, being uniform and suffering no variation, presents the appearance of a spiritual existence which is never anything but the same and uniform."

The *system enharmonic* was, as already stated,

superseded by the Pythagorean system, and when attempts were made to reinstate it, the ignorance of the mode of reckoning the enharmonic interval prevented a return to it.   Complaints about the loss music thus sustained are freely indulged in by Greek philosophers.   Plutarch is eloquent on the subject. He says (*De Musica*, vol. iii.)—

"The men of the present day have disdained all connection with the first of the systems, to the extent that with the main part of them there is not so much as the accidental recognition of the *enharmonic*, and so careless and indisposed to exertion are they upon the subject, as definitely to determine that the enharmonic diesis has left no trace of what it was in anything that falls within the power of comprehension."

Enharmonic tuning, or the system of tuning afresh for every change in the mode or genera, was however insisted upon by the Greeks, although the ratios for the true *enharmonic scale* once lost were never recovered.

It was the attempt to violate the great principle of enharmonic or perfect tuning which gave rise to the curious edict issued against Timotheus the Milesian, for appearing at the games of the Eleusian Ceres with a harp of eleven strings.   It must not be implied by the edict that harps with more than seven strings were objected to, for there is good

reason to believe that as scales were extended to
two octaves, instruments with a compass of two
octaves might have been allowed, but the offence
was, introducing eleven strings *within the octave* in
order to avoid retuning, when the mode or genera
was changed.   This plan involved a departure from
enharmonic or correct tuning, and substituted for it
a kind of tempered system.   Hence the severity of the
edict preserved in the Oxford edition of Aratus in
the Library of the British Museum, a translation of
which runs thus :—

" Whereas Timotheus the Milesian, after coming
into our state, holding in contempt the ancient music
and turning away from the harp practice with the
seven strings, introducing a multiplicity of sounds, is
spoiling the ears of youth, and with the multiplication of strings and ignoble and artificial novelty of
musical construction is dressing up music into something contrary to the pure and regulated kind,
working the degradation of musical construction into
chromatic instead of the enharmonic (with its accurate tuning of each string to the single sound
required from it), making a convertible (string) to be
applied to different purposes in turn.   And, moreover, being invited to take part in the games of the
Eleusinian Ceres did give loose to unseemly proceed-
ings which were a degradation to the story under
representation, which was the ' Sorrows of Semele,'
for the performance of which he has young persons
to teach.   It has seemed good for these reasons to

declarè that the kings and orators do reprimand
Timotheus, and further do enforce him to cut off
from the eleven-stringed lyre the superfluous
strings, leaving the seven tones.   To remove to the
utmost, annoyance to the state, that warning may
be taken against importing any of the improper customs into Sparta, and that the honour of the games
be not disturbed."

There is no proof nor inference that the pre-
Pythagorean Greeks inserted more than seven
notes within the octave upon which to set up
their scales or tetrachordal divisions, nor that
they gave to their titled scales attendant scales
on the fourth below.

It will be shown presently that the Greeks, after
Pythagoras's time inserted 12 sounds within the
octave, and gave the genera scales set up on
those sounds their hypo and hyper attendant
scales.*   But even the Pythagorean Greeks did not
give attendant scales on the fourth below, to their
prime and six evolved scales, but only to their
genera scales, to which they also gave attendant
scales on the fourth above.

REMARKS ON THE ANALYSIS OF THE OLD GREEK SCALES.

The ratios reckoned from the prime or first note
in all these analysed Greek scales very fairly exem-

---

* *Vide Alypius's " Genera Scales."—Edit Meibomius.*

plify the principle upon which the beautiful in
sound depends, viz., on ratios reckoned from a prime
of the simplest possible kind.   The ratios are—

Phrygian Mode	8 : 9	5 : 6	3 : 4	2 : 3	5 : 8	9 : 16
Dorian Mode	243 : 256	27 : 32	3 : 4	729 : 1024	81 : 128	9 : 16
Lydian Mode	9 : 10	4 : 5	3 : 4	27 : 40	3 : 5	9 : 16
Diatonic Genera Scale	15 : 16	27 : 32	3 : 4	45 : 64	81 : 128	9 : 16
Chromatic Genera Scale	15 : 16	9 : 10	3 : 4	45 : 64	27 : 40	9 : 16
Enharmonic Genera Scale	31 : 32	15 : 16	3 : 4	93 : 128	45 : 64	9 : 16

From the above results of our analysis it will be
seen how near were the old Greeks before the time
of Pythagoras in arriving at a correct exemplification
of the beautiful ; for though they did not formulate
a scale like our just *major* scale, their Phrygian
scale is *identical* with our *minor* descending scale,
when we use the minor seventh, ratio 9 : 16, in
place of ratio 5 : 9, as vocalists and violinists sometimes do.

Succeeding in establishing a minor scale so allied
to that we have adopted, it may well be regretted
that the old Greeks allowed, as we shall proceed
to show, the system of Pythagoras to supersede
their own tetrachordal divisions, and permitted the
influence of that great philosopher to blind them
to the fact that his method of exemplifying the
beautiful theory of simple ratios was the most
pernicious that could have well been devised.

CHAPTER IV

THE PYTHAGOREAN SERIES OF SOUNDS.

THE peculiar system which in its entirety may with certainty be attributed to Pythagoras (B.C. 540), is assumed to have been formulated by adopting intervals of perfect fifths, and then eliminating the octaves and arranging the sounds in the order of a scale. Thus, starting from a prime, say C, we have the following order—

$$C^{2:3} \mid G^{2:3} \mid D^{2:3} \mid A^{2:3} \mid E^{2:3} \mid B$$

Reducing D an octave, A an octave, E two octaves, and B two octaves, we have the series

C   D   E   G   A   B

To obtain the F, deficient in the above series, and fill up the wide interval between E and G, it appears that F as a fifth below the prime C was taken and raised an octave. The scale would thus be complete, and still the Pythagorean method be maintained of reckoning by perfect fifths. The ratios of these notes to the prime are 8 : 9, 64 : 81, 3 : 4, 2 : 3,

16 : 27, 128 : 243.   The Pythagorean scale consequently was as follows :—

THE PYTHAGOREAN SCALE.

	C	D	E	F	G	A	B	C
Intervening ratios ..	8 : 9	8 : 9	243 : 256	8 : 9	8 : 9	8 : 9	243 : 256	
Intervening cents ..	204	204	90	204	204	204	90	
Ratios from C	1	8 : 9	64 : 81	3 : 4	2 : 3	16 : 27	128 : 243	1 : 2
Cents from C	0	204	408	498	702	906	1110	1200

Although there is no doubt that the Chinese,
and perhaps other Oriental nations were accustomed to tune their instruments by fifths long
antecedent to the Pythagorean period, nevertheless,
as a mode of constructing a scale between a
prime and its octave, the originality of the above
scheme is attributed to Pythagoras.   It is moreover
an indisputable fact, that Pythagoras in whose
philosophy music held an important place, adopted
the system of reckoning by fifths, and assigned
reasons for so doing in such strict accordance with
other portions of his philosophy and teachings, that
his methods formed an era in the development of
Greek music which stamps them with paramount
interest and importance to all subsequent ages.

To apprehend the *raison d'etre* of a system, the
very errors of which have been cherished with
proscriptive reverence for its alleged author, it is
necessary briefly to recur to that portion of the
Pythagorean philosophy which selected special

numbers as sacred, because they corresponded with the elements, planetary order, and physiological structures.

The Chinese from time immemorial had contended for the sacredness of the number 5, because it corresponded to the five elements, the fingers of the hand, the senses, &c. The Egyptians had the most elaborate methods of defining the characteristics of special numbers, in which the 3rd, 4th, 5th, 7th, 10th, and 12th played conspicuous parts.*

There can be no question that the five-stringed lyre, the subdivision of a stretched string into four times four, and many other attempts to evolve numerical order from sounds, proceeded from the invariable tendency of ancient philosophers to connect the sciences with the laws of the visible universe.

The plastic Greek mind readily accepted and

---

* In Taylor's commentaries on the mysteries of Isis and Osiris there are many references to the Kabbalistic meanings attached to special numbers. Thus— the famous trinity of the Egyptians was derived from Osiris (the male), Isis (the female), and Horus (the child), or representation of created forms. The number *four* signified the perfect square, of which the four cardinal points were the representatives. *Five* was esteemed as a sacred number from the earliest antiquity ; the physiological reasons for this preference in the human structure being confirmed by the discovery of the five planets, the only number at that time known to the ancients. *Seven* was held sacred, as uniting the sacred numbers three and four. *Nine* was also a sacred number, as being the triad of three ; and *ten*, as filling up a numeral, from whence all further calculations must be repeated. The number *twelve* was held especially sacred, as being the triad of a perfect square, also as the final subdivision of the year mapped out by the twelve Zodiacal constellations. There can be but little doubt that the philosophers of Greece, who, like Orpheus, Pythagoras, Solon, Plato, and others, sat at the feet of the Egyptian Magians, to learn of "their wisdom," transferred many of their ideas to the various branches of science which were adopted and formulated by Grecian sages.

laboured to formulate these ideas, and although unquestionably borrowed from older nations, the genius of Grecian ideality threw the charm both of scientific development and æsthetic grace so completely around all that it adopted, that the fundamental idea is often lost sight of, in the improved method by which it has been unfolded. And thus it was that Pythagoras, the master mind of his time, brought his astronomical lore to bear on the science of sound. He asked, and professed to answer the question himself, "Why is consonance (*i.e.*, the beautiful in sound) determined by the ratio of small whole numbers?" The whole solution of this enigma Pythagoras found or professed to find in the order of the universe, where whole numbers and simple ratios prevail.

The correct numerical ratios existing between the seven tones of the Diatonic scale, according to Pythagoras, corresponded to the sun, moon, and five planets, and "the distances of the celestial bodies from the central fire," &c.

It was the elaboration of these figments of philosophy, and because the fifth as the central tone of the octave corresponded to the astronomical order in which the Samian sage ranged the sun and planets, that he laid such a deep stress upon a scale obtained from fifths only.

Even in the later and more advanced periods of Aristotle and Euclid, the same importance, pro-bably for the same cause, was attached to the fifth note from the prime of the scale.

It is a singular fact, whether it be considered in the light of design or coincidence, that the system of forming a scale by taking perfect fifths (the method of calculation introduced into Greece by Pythagoras) should correspond perfectly in its results with the old Dorian scale, if we commence the latter on the second note. For example, if we take the Dorian scale and compare it with the Pythagorean, calcu-lating the respective intervals by cents, we have the following results :—

DORIAN SCALE.

Intervening cents.. | B_{90} | C_{204} | D_{204} | E_{90} | F_{204} | G_{204} | A_{204} | B

Commencing the Pythagorean scale (produced by taking a series of perfect fifths and eliminating the octaves) and starting on C (the second note of the above-cited Dorian scale) we have the following similar series :—

PYTHAGOREAN SCALE.

Intervening cents.. | C_{204} | D_{204} | E_{90} | F_{204} | G_{204} | A_{204} | B_{90} | C

F

THE PYTHAGOREAN SERIES OF SCALES.

Having established the primal scale by a succession of fifths, as before shown, and called it by the title of the old Grecian tetrachordal scale, the *Lydian*, Pythagoras evolved from it, six other scales. These he obtained, not by setting up a fresh series of fifths on the several notes of his primal scale, but by making the second note of his first scale the prime of his second, and in this way constructing fresh scales upon each of the initial notes of the original series.

The six scales thus evolved were called

THE PHRYGIAN.
THE DORIAN.
THE HYPO-LYDIAN.
THE HYPO-PHRYGIAN.
THE HYPO-DORIAN.
THE MIXO-LYDIAN.*

TABLES OF PYTHAGOREAN SCALES.

THE PYTHAGOREAN LYDIAN.

	C	D	E	F	G	A	B	C
Intervening ratios..	$8:9$	$8:9$	$243:256$	$8:9$	$8:9$	$8:9$	$243:256$	
Intervening cents ..	204	204	90	204	204	204	90	
Ratios from C......	1	$8:9$	$64:81$	$3:4$	$2:3$	$16:27$	$128:243$	$1:2$
Cents from C	0	204	408	498	702	906	1110	1200

---

* The appropriation of these titles to scales quite different from those with which they were first associated, renders it necessary in alluding to them to distinguish them as the *Pythagorean* Lydian, *Pythagorean* Phrygian, &c., &c.

THE PYTHAGOREAN PHRYGIAN.

	D	E	F	G	A	B	C	D
Intervening ratios..	8:9	243:256	8:9	8:9	8:9	243:256	8:9	
Intervening cents ..	204	90	204	204	204	90	204	
Ratios from D......	1	8:9	27:32	3:4	2:3	16:27	9:16	1:2
Cents from D	0	204	294	498	702	906	996	1250

THE PYTHAGOREAN DORIC, OR DORIAN.

	E	F	G	A	B	C	D	E
Intervening ratios..	243:256	8:9	8:9	8:9	243:256	8:9	8:9	
Intervening cents ..	90	204	204	204	90	204	204	
Ratios from E......	1	243:256	27:32	3:4	2:3	81:128	9:16	1:2
Cents from E	0	90	294	498	702	792	996	1200

THE PYTHAGOREAN HYPO-LYDIAN.

	F	G	A	B	C	D	E	F
Intervening ratios..	8:9	8:9	8:9	243:256	8:9	8:9	243:256	
Intervening cents ..	204	204	204	90	204	204	90	
Ratios from F......	1	8:9	64:81	512:729	2:3	16:27	128:243	1:2
Cents from F	0	204	408	612	702	906	1110	1200

THE PYTHAGOREAN HYPO-PHRYGIAN.

	G	A	B	C	D	E	F	G
Intervening ratios..	8:9	8:9	243:256	8:9	8:9	243:256	8:9	
Intervening cents ..	204	204	90	204	204	90	204	
Ratios from G......	1	8:9	64:81	3:4	2:3	16:27	9:16	1:0
Cents from G	0	204	408	498	702	906	996	1202

THE PYTHAGOREAN HYPO-DORIAN.

	A	B	C	D	E	F	G	A
Intervening ratios..	8:9	243:256	8:9	8:9	243:256	8:9	8:9	
Intervening cents..	204	90	204	204	90	204	204	
Ratios from A......	1	8:9	27:32	3:4	2:3	81:128	9:16	1:
Cents from A	0	204	294	498	702	792	996	1200

THE PYTHAGOREAN MIXO-LYDIAN.

	B	C	D	E	F	G	A	B
Intervening ratios	243:256	8:9	8:9	243:256	8:9	8:9	8:9	
Intervening cents	90	204	204	90	204	204	204	
Ratios from B....	1	243:256	27:32	3:4	729:1024	81:128	9:16	1:2
Cents from B	0	90	294	498	588	792	996	1200

A comparison of the above scales with those of
the old Tetrachordal system, the names of which
Pythagoras adopted, will show that, despite the
similarity of the titles, the characteristics of all the

scales except the Dorian were totally dissimilar, and in place of the tetrachordal intervals, with all their intervening varieties, two only were used in the Pythagorean scales, namely—ratios 8 : 9, cents 204, and ratios 243 : 256, cents 90.

The question is, how far did these scales illustrate that principle of simple ratios or whole numbers which their author claimed to be essential for the production of "consonance"? We find that the ratios in respect to the prime in the seven Pythagorean scales were as follows :—

LYDIAN	1	8 : 9	64 : 81	3 : 4	2 : 3	16 : 27	128 : 243
PHRYGIAN	1	8 : 9	27 : 32	3 : 4	2 : 3	16 : 27	9 : 16
DORIAN	1	243 : 256	27 : 32	3 : 4	2 : 3	81 : 128	9 : 16
HYPO-LYDIAN...	1	8 : 9	64 : 81	512 : 729	2 : 3	16 : 27	128 : 243
HYPO-PHRYGIAN	1	8 : 9	64 : 81	3 : 4	2 : 3	16 : 27	9 : 16
HYPO-DORIAN ...	1	8 : 9	27 : 32	3 : 4	2 : 3	81 : 128	9 : 16
MIXO-LYDIAN...	1	243 : 256	27 : 32	3 : 4	729 : 1024	81 : 128	9 : 16

So that in the Pythagorean-Lydian mode there were only two simple ratios, viz., 3 : 4, 2 : 3, whilst in the old Grecian-Lydian there were three, viz., 4 : 5, 3 : 4, 3 : 5. In the Pythagorean-Phrygian mode there were two simple ratios, 3 : 4, 2 : 3, whilst in the old Grecian-Phrygian there were four, viz., 5 : 6, 3 : 4, 2 : 3, 5 : 8. The best feature of the Pythagorean system is the cccurrence of the perfect 5th in six of the model scales ; the worst is the use of ratios 64 : 81 (408 cents) and 16 : 27 (906 cents) for ratio 4 : 5 (386 cents) and ratio 3 : 5 (884 cents). Thus it will be seen that although the Pythagorean scale was founded upon the idea of exemplifying simple ratios, by the repetition of one single ratio, 2 : 3, and by the plan of evolving other scales out of the prime scale, several intervals were established which were in reference to the prime of the first scale, departures from the principle of simple ratios and divergences into complex ratios.

Referring to the fundamental idea of this treatise, *i.e.*, that a scale of tones determining the beautiful in sound must depend upon a succession of notes related to each other and a prime, by the simplest possible ratios, Pythagoras's method cannot be regarded as illustrating the principle in question. Pythagoras's mistake appears to have been, trying to blend his system of reckoning with his favourite idea of a correspondence between music and the cosmic order of the universe.

He failed to demonstrate this, as also the truth of his fundamental musical principle, that sounds to be beautiful in succession, must make consonances with each other—that is, be related in a manner that can be expressed by simple ratios.

REMARKS.

In studying the nature of the Greek scales, much confusion and not a little error has arisen from the indiscriminate use of the titles attached both to the old Greek and Pythagorean scales, for they are equally associated with the prefixes, Lydian, Dorian, and Phrygian. Unless, therefore, a clear definition of the author is given, from whom the scale is derived, or the system it represents, such as the old Greek, the Diatonic, Chromatic, &c., or the same titles with the prefix of the author's name, no correct idea can be formed of the series to which allusion is made.

To apprehend this more fully it may be as well to observe the differences existing between the subjoined diagrams :—

OLD GREEK LYDIAN CONJUNCT SCALE, WITH C AS THE INITIAL LETTER.

	C	D	E	F	G	A	B	C
Intervening ratios	9:10	8:9	15:16	9:10	8:9	15:16	♭8:9	
Intervening cents	182	204	112	182	204	112	204	
Ratios from C....	1	9:10	4:5	3:4	27:40	3:5	9:16	1:2
Cents from C	0	182	386	498	680	884	996	1200

THE PYTHAGOREAN LYDIAN SCALE.

	C	D	E	F	G	A	B	C
Intervening ratios	8:9	8:9	243:256	8:9	8:9	8:9	243:256	
Intervening cents	204	204	90	204	204	204	90	
Ratios from C....	1	8:9	64:81	3:4	2:3	16:27	128:243	1:2
Cents from C	0	204	408	498	702	906	1110	1200

THE OLD GEEK LYDIAN DIATONIC CONJUNCT GENERA SCALE.

	C	D	E	F	G	A	B	C
Intervening ratios	15:16	♭9:10	♭8:9	15:16	♭9:10	♭8:9	♭8:9	
Intervening cents	112	182	204	112	182	204	204	
Ratios from C....	1	15:16	27:32	3:4	45:64	81:128	9:16	1:2
Cents from C	0	112	294	498	610	792	996	1200

THE OLD GREEK LYDIAN CHROMATIC CONJUNCT GENERA SCALE.

	C	D♭	E♭♭	F	G♭	A♭♭	B♭♭	C
Intervening ratios		15:16	24:25	5:6	15:16	24:25	5:6	8:9
Intervening cents		112	70	316	112	70	316	204
Ratios from C....	1	15:16	9:10	3:4	45:64	27:40	9:16	1:2
Cents from C	0	112	182	498	610	680	996	1200

In addition to these four Lydian scales there is *Ptolemy's Lydian* (*vide* Ptolemy's scales).

CHAPTER V

ARCHYTAS' SERIES OF SOUNDS.

A RCHYTAS a philosopher who flourished about 400 B.C., although professing to be an adherent of the Pythagorean school in music, is yet considered to have differed from his great prototype so materially in the tetrachordal divisions of the genera scales, that he is credited by some commentators—with very little justice—as the originator of a new system. The truth is, the system of Archytas was to a certain extent that of the Pre-Pythagorean Greeks with slight modification. Our sources of information concerning the methods attributed to Archytas are derived from Ptolemy, who, in describing the scales of Archytas, refers to him as one of a notable group of five ancient musicians.

Helmholtz alludes to Archytas as the first ancient musician who reckoned the ratio 4 : 5 from the highest note of the tetrachord; but, as we have previously remarked, it would be more correct to credit him with being the first to formulate the plan of so doing, as the practice was common before the time of Archytas.

The ratios in the following diagrams are deduced from Ptolemy's "lengths of string."* The letters X, Y, Z, signify notes for which in our present system of notation we have no equivalents.

ARCHYTAS' ENHARMONIC GENERA SCALE.

	A	X	B	D	E	Y	F	a
Intervening ratios..	27 : 28	35 : 36	4 : 5	8 : 9	27 : 28	35 : 36	4 : 5	
Intervening cents ..	63	49	386	204	63	49	386	
Length of string ..	120	115·43	112·30	90	80	77·9	75	60
Ratios from D to A..	3 : 4	7 : 9	4 : 5	—	—	—	—	
Ratios from a to E..	—	—	—	—	3 : 4	7 : 9	4 : 5	1
Ratios from A	1	27 : 28	15 : 16	3 : 4	2 : 3	9 : 14	5 : 8	1 : 2
Cents from A	0	63	112	498	702	765	814	1200

ARCHYTAS' CHROMATIC GENERA SCALE.

	A	X	B	D	E	Y	F	a
Intervening ratios..	27 : 28	224 : 243	27 : 32	8 : 9	27 : 28	224 : 243	27 : 32	
Intervening cents ..	63	141	294	204	63	141	294	
Length of string....	120	115·43	106·40	90	80	77·9	71·7	60
Ratios from D to A..	3 : 4	7 : 9	27 : 32	1	—	—	—	
Ratios from a to E..	—	—	—	—	3 : 4	7 : 9	27 : 32	1
Ratios from A	1	27 : 28	8 : 9	3 : 4	2 : 3	9 : 14	16 : 27	1 : 2
Cents from A	0	63	204	498	702	765	906	1200

ARCHYTAS' DIATONIC GENERA SCALE.

	A	X	C	D	E	Y	G	a
Intervening ratios..	27 : 28	7 : 8	8 : 9	8 : 9	27 : 28	7 : 8	8 : 9	
Intervening cents ..	63	231	204	204	63	231	204	
Length of string....	120	105·43	101·15	90	80	77·9	67·30	60
Ratios from D to A..	3 : 4	7 : 9	8 : 9	1	—	—	—	
Ratios from a to E..	—	—	—	—	3 : 4	7 : 9	8 : 9	1
Ratios from A	1	27 : 28	27 : 32	3 : 4	2 : 3	9 : 14	9 : 16	1 : 2
Cents from A	0	63	294	498	702	765	996	1200

Our analysis of Archytas' scales shows that instead of setting up conjoint tetrachords he set up disjunct tetrachords, and so established the perfect fifth in all his genera scales.

---

* Claudii Ptolemæi Harmonicorum, Lib. II.

Archytas' ratios, according to Ptolemy, were—
For the *Enharmonic genera* scales :—
1, 27 : 28, 15 : 16, 3 : 4, 2 : 3, 9 : 14, 5 : 8, 1 : 2.
For the *Chromatic genera* scales :—
1, 27 : 28, 8 : 9, 3 : 4, 2 : 3, 9 : 14, 16 : 27, 1 : 2.
For the *Diatonic genera* scales :—
1, 27 : 28, 27 : 32, 3 : 4, 2 : 3, 9 : 14, 9 : 16, 1 : 2.

In pursuance of our analysis we shall now consider, with some attention the system of Aristoxenes.

CHAPTER VI

ARISTOXENES' SERIES OF SOUNDS.

IN connection with the history of Greek music, Aristoxenes holds a prominent and peculiar position, being the only eminent musical writer who openly combatted the opinions of Pythagoras on the question of how far mathematical proportions were to be regarded as the sole arbiters of musical scales to the exclusion of the ear, or what might now be designated æsthetic taste.

It is generally believed that Aristoxenes' main doctrine consisted in the plea, that the ear of the cultured listener should always be consulted in the arrangement of musical sounds.

It is also affirmed that he ignored the divisions of the tetrachord when derived only from mathematical theorems, alleging that the fourth, fifth, and eighth, being satisfactory concords, should govern the relations of all the other notes of the scale—in a word, that by carefully noting the difference between the fourth and fifth, and then, by adding or subtracting that interval to or from the concords, all the other required intervals could be so determined, as to fill up the scale. Of course the formation of a scale on such a plan as the above, excluding altogether any scientific plan of filling up the intervals, must be measurably decided by the taste or ear.

Assuming that the ear was sufficiently educated to determine the proper relative pitch of intervals Aristoxenes would have had instruments tuned to the sounds of

$$ \text{C} \qquad \text{F} \qquad \text{G} \qquad \text{C} $$

or notes that would furnish a prime, perfect fourth, perfect fifth, and octave. Making these notes the fundamentals of his system, then adding or selecting an interval equal to that existing between F and G, and placing notes at that interval above and below C, he would have D and B♭. On the same plan, by establishing a tone equal to the interval between F and G below F he would have E♭, and in like manner a note above G would give A. Thus the scale would be—

$$ \text{B}^{\flat 204} \mid \text{C}^{204} \mid \text{D}^{90} \mid \text{E}^{\flat 204} \mid \text{F}^{204} \mid \text{G}^{204} \mid \text{A} $$

The first evolved scale would be—

$$ \text{C}^{204} \mid \text{D}^{90} \mid \text{E}^{\flat 204} \mid \text{F}^{204} \mid \text{G}^{204} \mid \text{A}^{90} \mid \text{B}^{\flat} $$

By comparing these scales with those of Pythagoras it will be seen that the intervening intervals of the first, coincide with the Pythagorean Lydian, those of the second, with the Pythagorean Phrygian scale.

Of course the Pythagoreans who were the avowed opponents to the doctrines of Aristoxenes could only censure the *mode of evolving* their rival's scales. The results, as far as the succession of tones went, were too well in accord with their great master's scales to be rejected. The real point at issue seems to have been however, the necessity of Aristoxenes determining the relative pitch of his intervals by the ear—a standard, as all must admit, liable to most serious divergencies, whilst the Pythagoreans obtained their results from calculations based upon unimpeachable mathematical proportions.

But when Aristoxenes desired to illustrate his views through the genera scales, he found the necessity of obtaining smaller intervals than a tone, and this he aimed to arrive at, it is said, by dividing the whole tone by 2 or other desired numbers. This, Euclid * shows to be impossible : for to find two equal ratios whose product is equal to the ratio 8 : 9 necessitates finding the square root of 8, which is an irrational or *surd* quantity.

Again, the Pythagoreans alleged that the methods advocated by Aristoxenes were inconsistent, and that the attempt to reduce them to practical results would show them to be far more complicated than those of Pythagoras.

* Sectio Canonis.

We gather from Ptolemy's diagrams of Aristoxenes' scales, that the latter's plan of evolution was to regard the perfect fourth as divisible into thirty parts.

In the enharmonic scale he gives twenty-four parts to the first interval below the highest note of the tetrachord, three parts to the second, and three parts to the third.

Ptolemy does not give the ratios for these parts, but the lengths of the strings which produced the divisions, and these are

Highest note	C	60
„	A♭	76
„	W*	78
„	G	80
„	F	90
„	D♭	114
„	X*	117
Lowest note	C	120

so that the following was

ARISTOXENES' ENHARMONIC SCALE.

	C	W	D	F	G	X	A	C
Parts	3	3	24		3	3	24	
Intervening ratios	39:40	38:39	15:19		39:40	38:39	15:19	
Intervening cents	44	45	409	—	44	45	409	—
Ptolemy's length of string	120	117	114	90	80	78	76	60
Ratios from F to C	3:4	10:13	15:19	1	—	—	—	—
Ratios from c to G	—	—	—	—	3:4	10:13	15:19	1
Ratios from C	1	39:40	19:20	3:4	2:3	13:20	19:30	1:2
Cents from C	0	44	89	498	702	746	791	1200

* Notes for which we have no equivalent musical letters.

ARISTOXENES' CHROMATIC GENERA SCALES.

These are quoted by Ptolemy as three in number, and are styled the *Mollis Chromatica*, the *Sesquialterius Chromatica*, and the *Tonici Chromatica*.

THE MOLLIS CHROMATICA SCALE.

In this the tetrachords are divided into 22 parts, 4 parts, and 4 parts = 30 parts.

	$C^{4}_{29:30}$	$W^{4}_{28:29}$	$D^{22}_{45:56}$	$F^{-}_{8:9}$	$G^{4}_{29:30}$	$X^{4}_{28:29}$	$A^{22}_{45:56}$	c
Parts / Intervening ratios								
Intervening cents	59	60	379	204	59	60	379	—
Length of string	120	116	112	90	80	77·20	74·40	60
Ratios from F to C	3:4	45:58	45:56	1	—	—	—	—
Ratios from c to F	—	—	—	—	3:4	45:58	45:56	1
Ratios from C....	1	29:30	14:15	3:4	2:3	29:45	28:45	1:2
Cents from C	0	59	119	498	702	761	821	1200

SESQUIALTERIUS CHROMATICA SCALE.

In this the tetrachords are divided into 21, 4½, and 4½ parts = 30 parts.

	$C^{4\frac{1}{2}}_{77:80}$	$W^{4\frac{1}{2}}_{74:77}$	$X^{21}_{30:37}$	$F^{-}_{8:9}$	$G^{4\frac{1}{2}}_{77:80}$	$Y^{4\frac{1}{2}}_{74:77}$	$Z^{21}_{30:37}$	c
Parts / Intervening ratios								
Intervening cents	65	70	363	204	65	70	363	—
Length of string	120	115·30	111	90	80	77	74	60
Ratios from F to C	3:4	60:77	30:37	1	—	—	—	—
Ratios from c to G	—	—	—	—	3:4	60:77	30:37	1
Ratios from C....	1	77:80	37:40	3:4	2:3	77:120	37:60	1:2
Cents from C	—	65	135	498	702	767	837	1200

THE TONICI CHROMATICA SCALE.

In this the tetrachords are divided into 18, 6, and 6 parts = 30 parts.

	$C^{6}_{19:20}$	$W^{6}_{18:19}$	$D^{18}_{5:6}$	F	$G^{6}_{19:20}$	$X^{6}_{18:19}$	$A^{18}_{5:6}$	c
Parts / Intervening ratios								
Intervening cents	89	93	316	—	89	93	316	—.
Length of string	120	114	108	90	80	76	72	60
Ratios from F to C	3:4	15:19	5:6	1	—	—	—	—
Ratios from c to G	—	—	—	—	3:4	15:19	5:6	1
Ratios from C....	—	19:20	9:10	3:4	2:3	19:30	3:5	1:2
Cents from C	0	89	182	498	702	791	884	1200

ARISTOXENES' DIATONIC SCALES.

These scales are two in number, and are entitled the *Mollis Diatonica* and the *Intensii Diatonica.*

The Mollis Diatonica Scale,

in which the tetrachords are divided into 15, 9, and 6 parts = 30.

	C $_{19:20}^{6}$	D $_{35:38}^{2\ 9}$	D $_{6:7}^{\sharp15}$	F	G $_{19:20}^{6}$	A $_{35:38}^{\flat9}$	A $_{6:7}^{\sharp15}$	c
Parts								
Intervening ratios......								
Intervening cents	89	142	267	—	89	142	267	—
Length of string	120	114	105	90	80	76	70	60
Ratios from F to C	3 : 4	15 : 19	6 : 7	1	—	—	—	—
Ratios from e to G	—	—	—	—	3 : 4	15 : 19	6 : 7	1
Ratios from C..........	1	19 : 20	7 : 8	3 : 4	2 : 3	19 : 30	7 : 12	1 : 2
Cents from C	0	89	231	498	702	791	933	1200

THE INTENSII DIATONICA SCALE,

in which the tetrachords are divided into 12, 12, and 6 parts = 30.

	C $_{17:18}^{6}$	W $_{17:18}^{12}$	E $_{15:17}^{\flat12}$	F	G $_{19:20}^{6}$	X $_{17:18}^{12}$	B $_{15:17}^{\flat12}$	c
Parts								
Intervening ratios..								
Intervening cents ..	89	103	216	—	89	103	216	—
Length of string ..	120	114	102	90	80	76	68	60
Ratios from F to C ..	3 : 4	15 : 19	15 : 17	1	—	—	—	—
Ratios from e to F..	—	—	—	—	3 : 4	15 : 19	15 : 17	1
Ratios from C......	1	19 : 20	17 : 20	3 : 4	2 : 3	19 : 30	17 : 30	1 : 2
Cents from C	0	89	282	498	702	791	984	1200

The foregoing analyses of the divisions of the genera scale, attributed by Ptolemy to Aristoxenes, show, that in adopting what the latter claimed to be a less complex mode, viz., that of dividing the tetrachord into 30 parts, he fell into the error of establishing sounds at relationships which are less satisfactory to the ear than those of the method he aimed at superseding.

In fact, Aristoxenes' plan may be criticised as a simple mode of tuning out of tune.   Considering the basic claim of Aristoxenes' system—namely, that of making *the ear* the chief arbiter of musical order, rather than deferring the art solely to mathematical principles—the above-cited arrangement of the genera scales suffices to prove how faulty the system was.   Indeed, it is surprising that Greek sensitiveness could endure such a violation of the beautiful in sound as the substitution of the ratio $45 : 56$ (379 cents), as in the *Mollis Chromatica*, in place of so simple a ratio as $4 : 5$ (386 cents), with which the ear was well acquainted.

The opponents of Aristoxenes (and they were very numerous) were accustomed to say, "He might have been a good musician, but he was undoubtedly a bad mathematician."   They might have added with equal truth that his endeavour to appeal to the ear was as little calculated to satisfy æsthetic taste as to conciliate the judgment of mathematicians.

The chief advantage in Aristoxenes' mode of dividing the tetrachord in the seven scales, consists in the facility it affords for at once perceiving the relationship existing between intervals.   By the ratio system it cannot be determined at a glance

G

whether ratio 64 : 81 represents wider or closer intervals than ratio 4 : 5, whereas, in stating the number of parts into which the tetrachord is divided, *any given number* will at once determine the relationship of the parts.

In their dispute with the followers of Aristoxenes, the Pythagoreans announced a truth which no mere imaginative theory can displace—namely, that to satisfy the highest demands of the sensitive ear, the succession of sounds in musical order must be based upon mathematical divisions.

The theorem in itself is indisputable, nevertheless the Pythagoreans were no more infallible in its practical application than were the early Greeks. By adopting a mode for obtaining a succession of sounds closely adhering to an inapplicable theory, and wholly excluding the ear from any possible correction of that theory, they only succeeded in demonstrating their basic idea, without satisfying the demands of æsthetic taste.   On the other hand, the followers of Aristoxenes, making the ear the standard of what the order of succession should be, in establishing their prime scale by taking as the measure the intervals between the fourth and fifth for a *standard of audition*, and in overlooking the immense diversity of tastes, and the inexpediency of making the ear, however cultivated, the only arbiter of a proper interval, attempted the impossible task of separating the *science* from the art of music, and placing at the mercy of a fallible judge.

Not that Aristoxenes' system was destitute of rule or order, but the rule was arbitrary with himself alone and the order was destitute of a scientific basis.

Why both parties did not adopt the safe ground which nature supplied—namely, that of forming a scale out of such tones as would be given off by a mathematically-divided string, and then consulting the ear in the arrangement of the sounds—it boots not now to inquire.

Error is only the perversion of truth, and truth itself is often mistaken for falsehood, when it is *simply obscured* by error.   And thus it was that the Greek musicians lost the clue almost within their grasp, by which they might have discovered the basis of the beautiful in sound, and that most commonly, for the sake of maintaining certain dogmatic opinions.

The mathematicians disdained to consult the ear : whilst the theories of Aristoxenes determined him to abide only by the arbitration of the ear.   Had both methods been resorted to as factors in the evolution of music, the attainment of unity in æsthetic taste and mathematical science might have dated from Aristoxenes' time, instead of a distant period of two thousand years in the future.

CHAPTER VII

EUCLID'S MATHEMATICAL DIVISIONS OF A STRING,
AND RESULTING SERIES OF SOUNDS.

THE divisions of a stretched string, or "monochord," constituting the musical *Canon*, and
associated with the name of the most renowned
geometrician of ancient times, Euclid of Miletus
(B.C. 300), and the contemporary of Aristotle, will
now be analyzed, and the mode of procedure
explained. This is done in Euclid's "*19 Theorem,
Sectio Canonis.*" It is more readable in Col. Perronet
Thompson's Appendix to "*Just Intonation.*" We
therefore prefer to quote the latter.

"The string experimented upon is called Proslambanomenos, which may be illustrated by the
letter A (No. 1 of diagram). This string A was first
divided into four parts."

Three parts were taken, and the perfect fourth
established ratio 3 : 4. No. 2.

Two parts were taken, and the sound of the
octave A established. No. 3.

One part was taken, and the sound of the double
octave A^1 was given. No. 4.

The next experiment was to divide the length
which produced the fourth of the prime into two

parts, when the sound, the octave to D, was established.   No. 5.

Proslambanomenos was then divided into two equal parts, and one of these being again divided into three parts, two parts were taken and the octave of the fifth was established.   No. 6.

The length producing the octave of the fifth being doubled, a perfect fifth to the prime tone was established.   No. 7.

To apply the plan of simple ratios in determining the scale between the prime note and the fourth, a third part of the string producing the fifth was cut off, and the octave to the major second of the prime tone was thus obtained.   No. 8.

Then doubling the length producing the octave to the major second a tone an octave below was arrived at.   No. 9.

By dividing the length which produced the double octave of the prime into eight parts, and adding a length equal to one part, the grave minor seventh in the octave below was produced.   No. 10.

Then dividing the string producing the grave tone below the double octave into eight parts, and adding a length equal to one part, there was produced the octave to the minor sixth.   No. 11.

Then dividing the string producing the octave to the minor sixth into three equal parts, and adding a length equal to one part, there was produced the minor third in the octave above.   No. .12.

Then dividing the string producing the octave to the minor third into two equal parts, and adding a length equal to one part, there was produced the minor sixth in the octave below.   No. 13.

Then adding to the string producing the minor sixth a length equal to the difference between it and the string which gave the minor third in the octave above, there was produced the minor third in the octave below.   No. 14.

Lastly, taking three-fourths of the string which produced the fourth, there was established the grave minor seventh.   No. 15.

Thus were established—

No. 1	2	3	4	5	6	7	8	9	10	1	12	13	14	15
A	D	A	A	D	E	E	B	B	G	F	C	F	C	G

which (brought by the reduction of octaves into the compass of two octaves) produced—

THE EUCLIDEAN SERIES OF SOUNDS.

	A		B		C		D		E		F		G		a
Intervening ratios..		$8:9$		$243:256$		$8:9$		$8:9$		$243:256$		$8:9$		$8:9$		
Intervening cents ..		204		90		204		204		90		204		204		
Ratios from A......	1		$8:9$		$27:32$		$3:4$		$2:3$		$81:128$		$9:16$		$1:2$
Cents from A	0		204		204		498		702		792		996		1200

(SERIES CONTINUED).

	b		c		d		e		f		g		a^1
Intervening ratios..		$243:256$		$8:9$		$8:9$		$243:256$		$8:9$		$8:9$		
Intervening cents ..		90		204		204		90		204		204		

The sounds B C D E F G a b c d e f g a^1 are those required for the seven Pythagorean scales. Euclid having arrived at these sounds by divisions of the monochord, the scales formed with them are also called Euclidean. We give the titles, the ratios and cents.

THE EUCLIDEAN MIXO-LYDIAN.

	B	C	D	E	F	G	a
Intervening ratios	243 : 256	8 : 9	8 : 9	243 : 256	8 : 9	8 : 9	
Intervening cents	90	204	204	90	204	204	
Ratios from B....	1	243 : 256	27 : 32	3 : 4	729 : 1024	81 : 128	9 : 16
Cents from B	0	90	294	498	588	792	996

THE EUCLIDEAN LYDIAN.

	C	D	E	F	G	a	b
Intervening ratios..	8 : 9	8 : 9	243 : 256	8 : 9	8 : 9	8 : 9	
Intervening cents ..	204	204	90	204	204	204	
Ratios from C......	1	8 : 9	64 : 81	3 : 4	2 : 3	16 : 27	128 : 243
Cents from C	0	204	408	498	702	906	1110

THE EUCLIDEAN PHRYGIAN.

	D	E	F	G	a	b	c
Intervening ratios..	8 : 9	243 : 256	8 : 9	8 : 9	8 : 9	243 : 256	
Intervening cents ..	204	90	204	204	204	90	
Ratios	1	8 : 9	27 : 32	3 : 4	2 : 3	16 : 27	9 : 16
Cents	0	204	294	498	702	906	996

THE EUCLIDEAN DORIAN.

	E	F	G	a	b	c	d
Intervening ratios.	243 : 256	8 : 9	8 : 9	8 : 9	243 : 256	8 : 9	
Intervening cents.	90	204	204	204	90	204	
Ratios.............	1	243 : 256	27 : 32	3 : 4	2 : 3	81 : 128	9 : 16
Cents	0	90	294	498	702	792	996

THE EUCLIDEAN HYPO-LYDIAN.

	F	G	A	D	c	d	e
Intervening ratios.	8 : 9	8 : 9	8 : 9	243 : 256	8 : 9	8 : 9	
Intervening cents.	204	204	204	90	204	204	
Ratios	1	8 : 9	64 : 81	512 : 729	2 : 3	16 : 27	125 : 243
Cents	0	204	408	612	702	906	1110

THE EUCLIDEAN HYPO-PHRYGIAN.

	G		a		b		c		d		e		f
Intervening ratios.		8 : 9		8 : 9		243 : 256		8 : 9		8 : 9		243 : 256	
Intervening cents.		204		204		90		204		204		90	
Ratios	1		8 : 9		64 : 81		3 : 4		2 : 3		16 : 27		9 : 16
Cents	0		204		408		498		702		906		996

THE EUCLIDEAN LOCRIAN.

	a		b		c		d		e		f		g
Intervening ratios.		8 : 9		243 : 256		8 : 9		8 : 9		243 : 256		8 : 9	
Intervening cents.		204		90		204		204		90		204	
Ratios	1		8 : 9		27 : 32		3 : 4		2 : 3		81 : 128		9 : 16
Cents	0		204		294		498		702		792		996

Meibomius[*] gives the above lettering and titles, and makes it clear that Euclid's Mixo-Lydian was below the Lydian by placing against the initial note of the Mixo-Lydian, viz. B, the term *hypate hypaton, i.e.,* the note next to the uppermost, the uppermost being the lowest in the pitch according to the old Greek plan of description. He also determines the relationships of all the initial sounds to Proslambanomenos, by placing

Against the initial C the term Parapate Hypaton.

,,   ,,   ,,   D   ,,   Lichanos Hypaton.

,,   ,,   ,,   E   ,,   Hypate Meson.

,,   ,,   ,,   F   ,,   Parapate Meson.

,,   ,,   ,,   G   ,,   Lichanos Meson.

,,   ,,   ,,   a   ,,   Mese.

Meibomius also shows that the scale on *mese* (a) was the scale of Proslambanomenos A by writing *mese vel Proslambanomenos,* and he styles the scale on *a* the *Locrian vel communis,* meaning thereby

---

[*] Marci Meibomii notæ in Euclid's introd. Harmon.

the scale when A is the initial, be it on Proslambanomenos or on the note next above the initial of the Hypo-Phrygian.

It will be observed that Euclid's series of sounds comprises four tetrachords and one whole tone. This is shown in the following diagram, in which the highest note of the series is the highest note of the first tetrachord, and the tetrachords are formed below it.   (Diagram read from right to left).

Note over.	Fourth Tetrachord.	Third Tetrachord.	Second Tetrachord.	First Tetrachord.
A |	B C D E	F G a	b c d e	f g a^1

In consequence of the circumstance that the Euclidean divisions necessarily range in tetrachords, many musicians allude to Euclid's series of sounds as if it were established by tetrachords, and describe A as an *added note*, disregarding the plain language of Euclid that A is *Proslambanomenos* the *string taken to begin with* or "experimented upon." The practice of designating *Proslambanomenos* the *added note* has caused great confusion and many errors.

Euclid's series certainly comprises four tetrachords and one whole tone, but we might as well call the remainder of a line of 37 inches when reduced to feet an *added inch*, as what is over when four tetrachords are taken out of two octaves "an *added note*."

There is no word used by Euclid which gives a pretext for using the term *added note*. To show this, we append a fac-simile of Euclid's divided string, with the titles of the different parts.   No other words are inscribed on this diagram, and the tetrachords marked out by Helmholtz are not in the Euclidean diagram handed down by Meibomius, though they are alluded to in Euclid's *Int_i oduction to Ha_i mony*, in the instructions how to set up the *genera scales*.

EUCLID'S CANON.*

B

Nete Hyperbolæon — E
M — Paranete Hyperbolæon
N — Trite Hyperbolæon
Nete Diezeugmenon — H
Nete Synemmenon — Z
X — Trite Diezeugmenon
Paramese — K
— *Trite Synemmenon* †
Mese — D

R — Meson Diatonos

O — Parypate Meson

Hypate Meson — ☉

G — Hypaton Diatonos

P — Parypate Hypaton

Hypate Gravis — L

Proslambanomenos — Λ

---

* The letters of the "Canon" refer only to Euclid's text.
† The division *Trite Synemmenon*, which gives *bb*, is not alluded to in Euclid's 19th Theorem, but it is referred to in his 20th Theorem.

EXPLANATION OF THE GREEK TITLES OF EUCLID'S DIVISIONS.

*
A   Proslambanomenos ......The string taken to begin with.
B   Hypate Gravis............The uppermost, *i.e.*, the note next to
                        Proslambanomenos.
C   Parapate Hypaton ......The next to the uppermost.
D   Hypaton Diatonis ......The diatone, *i.e.*, the tone lying
                        between two whole tones.
E   Hypate Meson   ....The uppermost, next to the diatone.
F   Parapate Meson ....The uppermost but one.
G   Meson Diatonis .. ..The middle diatone.
a   Mese ....The middle of the string.
bb Trite Synemmenon ......The third sound of the third tetrachord
b   Paramese ....The next to the middle of the string.
c   Trite Diezeugmenon ...The third sound of the second
                        tetrachord.
d   Nete Synemmenon ......The undermost but one.
e   Nete Diezeugmenon ...The undermost.
f   Trite Hyperbolœan ......The third sound of the first tetrachord.
g   Parauete Hyperbolæon...The undermost but one.
*a*¹ Nete Hyperbolæon ......The undermost.

EUCLID'S OBJECT IN DIVIDING THE MONOCHORD.

Euclid established no new scales by his divisions. He appears to have made them with a view of showing that however fanciful was Pythagoras' idea of deriving his succession of sounds from a series of perfect fifths, his succession had a mathematical basis, *i.e.*, could be deduced from certain mathematical divisions of a string. Perhaps, at the same time, he desired to establish a less circuitous mode of arriving at Pythagoras' results than was adopted by

---

* Letters which will be used as equivalents of the Greek titles.

that philosopher.   Certain it is that Euclid effected
no improvement on the Pythagorean intervals.   He
simply adopted the order of sounds bequeathed by
Pythagoras, and showed that their propriety could
be referred to the fact that they agreed with
certain divisions of a string.   Had Euclid formulated
the series of sounds that goes by his name (besides
that of Pythagoras) as the embodiment of his special
idea of what a scale should be, he would have left us
a problem worthy of his own solution, namely, why
he did not adopt after the ratios 1 : 2, 2 : 3, 3 : 4, the
next most simple ratios of 4 : 5.   But Euclid exercised his great powers not in formulating a series of
sounds but in proving that the mathematicians were
justified in accepting the Pythagorean series, and
that Aristoxenes' disciples were wrong in slighting it.

CHAPTER VIII

THE EFFECT OF THE PYTHAGOREAN SYSTEM ON THE GENERA SCALES.

THE Pythagorean system, besides completely super-seding the old Grecian modes, and appropriating their titles to the new order of scales derived from the one primal series set up by Pythagoras, con-siderably affected the *Genera scales*—that is, the three tetrachordal series included under the appella-tions of the Diatonic, Chromatic, and Enharmonic.

The Pythagorean Greeks it appears, adopted the tetrachordal system, but not the old Greek ratios in formulating these genera scales; for the instructions in Euclid's *Introduction to Harmony* are to the effect that the sounds in the diatonic genera scales must move from *acute to grave, i.e.,* from the highest to the lowest note of the *tetrachord*, by a *whole tone,* a *whole tone,* and a *half tone,** or from *grave to acute, i.e.,* from the lowest note of the tetrachord towards the highest, by a *half tone,* a *whole tone,* and a *whole tone.* In the *Introduction to Harmony* the names are given of the sounds of the tetrachords with which

---

* Per Tonum, Tonum, Hemitonium. Euclid.

the canonical divisions were associated.* From these it will be seen that the "canon," besides providing for the Pythagorean series of sounds (comprised in the modes) provided for $b\flat$ as well as $b\natural$ in the second octave from A (Proslambanomenos), and that the note ($b\flat$) enabled the Pythagorean Greeks to mark off five tetrachords out of the Euclidean series.

We give the series, and indicate by ⌢ the tetrachords. Read from right to left :—

The letters B C D E F G a in the first octave are equivalents for the Greek terms. ‡ $b\flat$ is *Trite*

---

* No doubt prevails as to the authenticity of Euclid's Sectio. Canonis, but there are good reasons for believing that the *Introduction to Harmony* is not the work of the great mathematician.

† Certain notes were standing notes, indicated thus, St. These did not alter in the different genera scales.

‡ B equivalent for Hypate Hypaton in Euclid's Introd.   In the division of the canon the title is Hypate Gravis.

C equivalent for Parypate Hypaton in Euclid's Introd.   In the division of the canon the title is the same.

D equivalent for Lichanos Hypaton Diatonos in Euclid's Introd.   In the division of the canon the title is Hypaton Diatonos.

E equivalent for Hypate Meson in Euclid's Introd.   In the division of the canon the title is the same.

F equivalent for Parypate Meson in Euclid's Introd.   In the division of the canon the title is the same.

G equivalent for Lichanos Meson Diatonos in Euclid's Introd.   In the division of the canon the title is Meson Diatonos.

Synemmenon; *b* with a natural after, thus *b♮*, in the second octave, is *Paramese*; *c* in the third tetrachord is *Paranete Synemmenon Diatonos*, but in the second tetrachord it is *Trite Diezeugmenon*, though the same note in sound: *d* in the third tetrachord is *Nete Synemmenon*, in the second tetrachord it is *Paranete Diezeugmenon Diatonos*, though the same note in sound. For the other four sounds we give our equivalents *e f g a*. The series was—

B C D E F G *a bb c d b♮ c d e f g a¹*

Thus the Pythagorean Diatonic tetrachords, five in number, are shown, as also the complete *Diatonic genera scale*. Proslambanomenos is not included in the tetrachords, as it was the note and string from which the series was derived. It was however always placed before *Hypate Hypaton*, to indicate the base of the series.

---

a equivalent for Mese in Euclid's Introd. In the division of the canon the title is the same.

bb equivalent for Trite Synemmenon in Euclid's Introd.

b equivalent for Paramese in Euclid's Introd. In the division of the canon the title is Paramese.

c equivalent for titles as stated in text. In the division of the canon the title is Trite Diezeugmenon.

d equivalent for titles as stated in text. In the division of the canon the title is Nete Synemmenon.

e equivalent for Nete Diezeugmenon in Euclid's Introd. In the division of the canon the title is the same.

f equivalent for Trite Hyperbolæon in Euclid's Introd. In the division of the canon the title is the same.

g equivalent for Paranete Hyperbolæon Diatonos in Euclid's Introd. In the division of the canon the title is Paranete Hyperbolæan.

a¹ equivalent for Nete Hyperbolæan in Euclid's Introd. In the division of the canon the title is the same.

THE PYTHAGOREAN CHROMATIC GENERA SCALE.

The instructions for setting up this scale are also given in Euclid's *Introduction to Harmony*. They are " move from *acute to grave* (*i.e.* in the tetrachordal divisions already alluded to) by a *minor third** (ratio 5 : 6, cents 316), by a *half tone*, a *half tone*, or from *grave to acute* by a *half tone*, a *half tone* and a *minor third*. So that the Chromatic genera tetrachords were

5th Tetra.		4th Tetra.		3rd Tetra.			
St.		St.		St.			St.
B_{91}^{243}	C_{91}^{243}	$D\flat_{316}^{5:6}$	E_{91}^{243}	F_{91}^{243}	$G\flat_{316}^{5:6}$ a $b\flat$ $c\flat$		d

2nd Tetra.		1st Tetra.		
St.		St.		St.
$b\natural$ c $d\flat$		e f $g\flat$		a

Proslambanomenos was placed (as in the Diatonic) as the base of the series of notes comprising the chromatic intervals. To make the chromatic intervals, D of the diatonic was lowered to D♭, and G of the diatonic was lowered to G♭. *c* in the third tetrachord was lowered to *c♭*, *d* in the second tetrachord was lowered to *d♭*, and *g* in the first tetrachord was lowered to *g♭*. Thus the series was

B C D♭ E F G♭ *a b♭ c♭ d b♮ c d♭ e f g♭ a*[1]

---

* Per Triemitonium, Hemitonium, and Hemitonium. Euclid.

H

D of the diatonic lowered to D♭, was called Lichanos Hypaton Chromatice.

G of the diatonic lowered to G♭ was called Lichanos Meson Chromatice.

c of the third tetrachord lowered to c♭ was called Paranete Synemmenon Chromatice.

d of the second tetrachord lowered to d♭ was called Paranete Diezeugmenon Chromatice.

g of the first tetrachord lowered to g♭ was called Paranete Hyperbolæon Chromatice.

The other sounds retained in the chromatic tetrachord preserved the names given to them in the diatonic genera tetrachords. Some of these were *"stantes"* or immovable notes in all the genera scales.

THE PYTHAGOREAN ENHARMONIC GENERA SCALE.

The instructions for setting up this scale in the work alluded to are—move from acute to grave by ditonum, an interval of two whole tones (ratio 64 : 81, cents 408), and by diesin (45 cents), and diesin (45 cents), or from grave to acute by *diesin,* *diesin,* and *ditonum,* so that the

ENHARMONIC TETRACHORDS WERE

St.   St.   St.   St.   St.   St.

B^diesis_45 D^bb diesis_45 C^64:81_408 E G^1bb F   a c♭♭ l♭ d b♮ d♭♭ c e g♭♭ f a¹ or differently named the series was

$$\overbrace{\text{B}\;\; \overset{*}{\text{W}}\;\; \text{C}}\;\; \overbrace{\text{E}\;\; \text{W}\;\; \text{F}} \left. \begin{array}{l} a \\ w \\ b\flat \\ d \end{array} \right] \;\; \overbrace{b \;\; w \;\; c} \;\; \overbrace{e \;\; w \;\; f} \;\; a$$

To make the enharmonic intervals

D of the diatonic was lowered to D♭♭,† and called Parypate Hypaton.

C was called Lichanos Hypaton Enarmonios.

G of the diatonic was lowered to G♭♭, and called Parypate Meson.

F was called Lichanos Meson Enarmonios.

c of the third diatonic tetrachord was lowered to c♭♭, and called Trite Synemmenon.

b♭ was called Paranete Synemmenon Enarmonios.

d of the second diatonic tetrachord was lowered to d♭♭, and called Trite Diezeugmenon.

c of the second Tetrachord was called Paranete Diezeugmenon Enarmonios.

g of the first diatonic tetrachord was lowered to g♭♭, and called Trite Hyperbolæon.

f of the first Tetrachord was called Paranete Hyperbolæon Enarmonios.

The other sounds retained the names given to them in the diatonic genera scales.

Thus do we show the diatonic, chromatic, and

---

* *W* being diesis.

† D♭♭ is here used as the symbol of a sound midway between B and C.

enharmonic tetrachords and the Pythagorean scales that were formulated with them.

According to Alypius,* the diatonic, chromatic, and enharmonic scales became transposed—that is, set up on key notes of higher pitch.

By considering as Proslambanomenos the Lydian C and reckoning from it, all the notes given to the genera scales in Euclid's Introduction to Harmony were set up in what Alypius describes the *Lydian mode*. Alypius's translated words are, "*The notes of the Lydian mode according to the diatonic, chromatic, and enharmonic genus.*"† To the Lydian mode as well as to the other modes, Alypius gives attendant modes on the fourth below and fourth above. The attendant mode on the fourth below he entitles the Hypo mode, that on the fourth above the Hyper mode.

The diatonic genera scales in the Lydian and the two attendant modes set up according to Euclid's formulæ were as follows :—

THE DIATONIC SCALE IN THE HYPO-LYDIAN MODE.

G | a bb c d eb f g | a♮ bb c d eb f g
     5th Tetra.   4th Tetra.   2nd Tetra.   1st Tetra.
                              ab
                              ♭b
                              c
                         3rd Tetra.

---

* Alypii "Introductio Musica." Alypius lived about A.D. 380.

† In other words, the notes of the genera scales appertaining to the Lydian modes.

THE DIATONIC IN THE LYDIAN MODE.*

C | *d*   *e♭*   *f*   *g*   *a♭*   *b♭*   { *c*   *d♮*   *e♭*   *f*   *g*   *a♭*   *b♭* ·   *c*

     5th Tetra.   4th Tetra.   { *d♭*   2nd Tetra.   1st Tetra.
                            { *e♭*
                            { *f*

                 3rd Tetra.

THE DIATONIC IN THE HYPER-LYDIAN MODE.

F | *g*   *a♭*   *b♭*   *c*   *d♭*   *e♭* { *f*   *g♮*   *a♭*   *b♭*   *c*   *d♭*   *e♭*   *f*
                              { *g♭*
                              { *a♭*
                              { *b♭*

Alypius shows that the Greeks set up their genera scales on a sound higher than C, by ratios 243 : 256, which we call C♯. They entitled the mode the Eolean. Alypius heads his description thus : *" The notes of the Eolean mode according to the diatonic genera."*

Placing Proslambanomenos on C♯, and giving it, its attendants, we have—

THE DIATONIC SCALE IN THE HYPO-EOLEAN MODE.

G♯ | *a♯*   *b*   *c♯*   *d♯*   *e*   *f♯* { *g♯*   *a♯*   *b*   *c♯*   *d♯*   *e*   *f♯*   *g♯*
                            { *a*
                            { *b*
                            { *c♯*

THE DIATONIC SCALE IN THE EOLEAN MODE.

C♯ | *d♯*   *e*   *f♯*   *g♯*   *a*   *b* { *c♯*   *d♯*   *e*   *f♯*   *g♯*   *a*   *b*   *c♯*
                            { *d♮*
                            { *f♯*

* The capital letters in the diagrams are not comprised in the scales, but show which note is Proslambanomenos, or the base of the series. The Diatonic in the Lydian Mode is *d e♭* (*g* and *e♭*) and not *c d e♭ f*, as Dr. Burney and other writers would imply.

THE DIATONIC SCALE IN THE HYPER-EOLEAN MODE.

$$F^\sharp \mid g^\sharp \quad a \quad b \quad c^\sharp \quad d \quad e \quad \begin{cases} f^\sharp \\ g^\natural \\ a \\ b \end{cases} \quad g^\sharp \quad a \quad b \quad c^\sharp \quad d \quad e \quad f^\sharp$$

5th Tetra. 4th Tetra. 3rd Tetra. 2nd Tetra. 1st Tetra.

The diatonic scales were then reckoned from the Phrygian D, and its attendants A and G.

THE DIATONIC SCALE IN THE HYPO-PHRYGIAN MODE.

$$A \mid b \quad c \quad d \quad e \quad f \quad g \begin{cases} a \\ bb \\ c \\ d \end{cases} \quad b\natural \quad c \quad d \quad e \quad f \quad g \quad a$$

THE DIATONIC SCALE IN THE PHRYGIAN MODE.

$$D \mid e \quad f \quad g \quad a \quad bb \quad c \begin{cases} d \\ eb \\ f \\ g \end{cases} \quad e\natural \quad f \quad g \quad a \quad bb \quad c \quad d$$

THE DIATONIC SCALE IN THE HYPER-PHRYGIAN MODE.

$$G \mid a \quad bb \quad c \quad d \quad eb \quad f \begin{cases} g \\ ab \\ bb \\ c \end{cases} \quad a\natural \quad bb \quad c \quad d \quad eb \quad f \quad g$$

On the note higher than D, by ratio 243 : 256, called the *Iastian*, and its attendants (the Hypo-Iastian and Hyper-Iastian), the genera scales were set up thus—

THE DIATONIC SCALE IN THE HYPO-IASTIAN MODE.

$$A^\sharp \mid b^\sharp \quad c^\sharp \quad d^\sharp \quad e^\sharp \quad f^\sharp \quad g^\sharp \begin{cases} a^\sharp \\ b \\ c^\sharp \\ d^\sharp \end{cases} \quad b^\sharp \quad c^\sharp \quad d^\sharp \quad e^\sharp \quad f^\sharp \quad g^\sharp \quad a^\sharp$$

THE DIATONIC SCALE IN THE IASTIAN MODE.

D♯ | c♯ f♯ g♯ a♯ b c♯ { d♯ e♮ f♯ g♯ } e♯ f♯ g♯ a♯ b c♯ d♯
5th Tetra.   4th Tetra.   2nd Tetra.   1st Tetra.

3rd Tetra.

THE DIATONIC SCALE IN THE HYPER-IASTIAN MODE.

G♯ | a♯ b c♯ d♯ e f♯ { g♯ a b c♯ } a♯ b c♯ d♯ e f♯ g♯

In the Pythagorean Dorian E mode, and on its attendants, B and A, the genera scales were—

THE DIATONIC SCALE IN THE HYPO-DORIAN MODE.

B | c♯ d e f♯ g a { b c♮ d e } c♯ d e f♯ g a b

THE DIATONIC SCALE IN THE DORIAN MODE.

E | f♯ g a b c d { e f♮ g a } f♯ g a b c d e

THE DIATONIC SCALE IN THE HYPER-DORIAN MODE.

A | b c d e f g { a b♭ c d } b♮ c d e f g a

THE PYTHAGOREAN CHROMATIC GENERA SCALES.*

In the same modes as those in which the Diatonic scales were set up, were also set up the chromatic genera scales according to the instructions in Euclid's *Introduction to Harmony*. It suffices to illustrate the chromatic genera scales in the Lydian and attendant modes, which were as follows:

CHROMATIC GENERA SCALE IN HYPO-LYDIAN MODE.

G | a bb cb d eb fb | g a♮ bb cb d eb fb g
       5th Tetra.   4th Tetra.   { ab   2nd Tetra.   1st Tetra.
                                     bbb
                                     c
                              3rd Tetra.

CHROMATIC GENERA SCALE IN LYDIAN MODE.

C | d eb fb g ab bbb | c d♮ eb fb g ab bbb c
                                   { db
                                     ebb
                                     f

CHROMATIC GENERA SCALE IN HYPER-LYDIAN MODE.

F | g ab bbb c db ebb | f g♮ ab bbb c db ebb f
                                    { gb
                                      abb
                                      bb

THE PYTHAGOREAN ENHARMONIC SCALES.*

In the same modes as those in which the Diatonic scales were set up were also set up the Enharmonic scales according to the instructions in Euclid's

---

* As given by Alypius.

Introduction to Harmony. The Enharmonic genera scales in the Lydian and attendant modes were as follows :—

ENHARMONIC GENERA SCALE IN HYPO-LYDIAN MODE.

G | a *w bb d w eb { g a♮ w bb d w eb g
              w
5th Tetra.   4th Tetra. { a b   2nd Tetra.   1st Tetra.
                          c
              3rd Tetra.

ENHARMONIC GENERA SCALE IN LYDIAN MODE.

C | d w eb g w ab { c d♮ w eb g w ab c
                          w
                          d b
                          f

ENHARMONIC GENERA SCALE IN HYPER-LYDIAN MODE.

F | g w ab c w db { f g♮ w ab c w db f
                          w
                          gb
                          bb

Alypius describes fifteen diatonic, fifteen chromatic, and fifteen enharmonic scales, but as three in each genus were merely octaves of three others, there were really only twelve scales in each genus.

---

* *w* is diesis from the notes it is placed between.

EUCLID'S MIXED SCALE.

In Euclid's Introduction allusion is made to a mixed scale—that is, a scale which included the characterising interval of all the scales, viz., the diatonic, the chromatic, and the enharmonic intervals. Such a scale in the Lydian mode would be

C | d w eb fb f♮ g w ab bbb bb ⎰c   d w ebb eb fb f♮ g w abb ab bbb bb c
       5th Tetra.   4th Tetra.   ⎱dbb   2nd Tetra.   1st Tetra.
                                        db
                                        ebb
                                        eb
                                        f
                        3rd Tetra.

Alypius does not allude to the mixed scale, and if practical use were ever made of it, it does not appear to have been generally adopted.

REMARKS.

Our analysis of the genera scales, as affected by the Pythagorean system, shows that, great as were the old Grecian departures from single ratios in these scales, Pythagoras's system made some of the ratios still more complex, whilst Euclid's division of the Canon which gives *Trite Synemmenon* extended or sanctioned the extension of the genera scales.

CHAPTER IX

THE SCALES OF ERATOSTHENES.

THE name of Eratosthenes, an ancient Greek philosopher, whose period is given at about 275 B.C., has become sufficiently prominent in the history of Greek musical art to justify an inquiry concerning the specialities of his teachings and experiments.

According to the account of Eratosthenes, given by Ptolemy in his *Harmonicorum,* he was one of five ancient Greek musicians who introduced some slight alterations in the intervals of the genera scales. These alterations are illustrated in the following diagrams. The lengths of string are given by Ptolemy. From these the ratios and cents are deduced.

ERATOSTHENES' ENHARMONIC SCALE.

	C	W	D	F	G	X	A	c
Intervening ratios..	39 : 40	38 : 39	215 : 19		39 : 40	38 : 39	215 : 19	
Intervening cents ..	44	45	409		44	45	45	
Length of string ..	120	117	114	90	80	78	76	60
Ratios from F to C..	3 : 4	10 : 13	15 : 19	1	—	—	—	--
Ratios from c to G ..	—	—	—	—	3 : 4	10 : 13	15 : 19	1
Ratios from C	1	39 : 40	19 : 20	3 : 4	2 : 3	26 : 40	81 : 128	1 : 2
Cents from C	0	44	89	498	702	746	791	1200

In this scale, the fifth, C to G is perfect and the ratio 15 : 19, cents 409, is nearly equivalent to the

Pythagorean major third, ratio 64 : 81, cents 408. Thus it is apparent that Eratosthenes was a follower of Pythagoras, that he adopted disjunct tetrachords, and so had the fifth of his scales perfect, whilst his major third was a slight variation without being any improvement upon his model.

ERATOSTHENES' CHROMATIC GENERA SCALE.

	C	W	D	F	G	X	A	c
Intervening ratios	19:20	18:19	5:6	8:9	19:20	18:19	5:6	
Intervening cents	88	94	316	204	88	94	316	
Length of string	120	114	108	90	80	76	72	60
Ratios from F to C	3:4	15:19	5:6	1	—	—	—	—
Ratios from c to G	—	—	—	—	3:4	15:19	5:6	1
Ratios from C	1	19:20	15:16	3:4	2:3	19:30	3:5	1:2
Cents from C	0	88	182	498	702	790	884	1200

In the above tetrachords the minor third is ratio 5 : 6, cents 316, as it is in the old Grecian chromatic and in the Pythagorean chromatic genera scales.

ERATOSTHENES' DIATONIC GENERA SCALE.

	C	D	E	F	G	A	B	c
Intervening ratios ..	243:256	♭8:9	♭8:9	8:9	243:256	♭8:9	♭8:9	
Intervening cents ..	90	204	204	204	90	204	204	
Length of string	120	113·54	101·15	90	80	75·56	67·30	60
Ratios from F to C ..	3:4	64:81	8:9	1	—	—	—	—
Ratios from c to G ..	—	—	—	—	3:4	64:81	8:9	1
Ratios from C	1	243:256	27:32	3:4	2:3	81:128	9:16	1:2
Cents from C	0	90	294	498	702	792	990	1200

This scale is formed with the Doric tetrachord, and therefore it will be seen that Eratosthenes' scales contribute nothing towards the results we seek to discover—namely, an improvement in the exemplification of simple ratios.

CHAPTER X

DIDYMUS' SCALES.

WE cite the scales of Didymus because they have been claimed by several authorities as approximating closely to our own just scale. The writings of this renowned Greek belong, in a general sense, to history.

His musical scales are quoted by Ptolemy, whose authority we rest upon for what follows.

The three genera scales of Didymus, as quoted by Ptolemy, are one enharmonic, one chromatic, and one diatonic.

DIDYMUS' ENHARMONIC SCALE.

	C	W	D♭	F	G	Y	A♭	c
Intervening ratios	31:32	30:31	24:5	8:9	31:32	30:31	24:5	
Intervening cents	55	57	386	204	55	57	386	
Length of string	120	117	114	90	80	77·30	75	60
Ratios from F to C	3:4	24:31	4:5	1	—	—	—	·—
Ratios from c to G	—	—	—	—	3:4	24:31	4:5	1
Ratios from C....	1	31:32	15:16	3:4	2:3	31:48	5:8	1:2
Cents from C....	0	55	112	498	702	757	814	1200

In the above tetrachordal divisions the ratio 4:5 between F and D♭ points to the old Grecian enharmonic tetrachord, for the ratio of the major third in the Pythagorean divisions is 64:81. Euclid says that the enharmonic moved from acute to grave by *ditonum, diesin and diesin, i.e.,* by an interval the width of *two major tones,* by a quarter tone and a quarter tone ; but Euclid alluded to the enharmonic scale as affected by the Pythagorean doctrine. Didymus ignored that doctrine, and exemplified the old Greek plan ; making the enharmonic move from acute to grave by an interval the width of *one major tone and one minor tone, i.e.,* by a major third, ratio 4:5, cents 386. The defect, the ratio 15:16, he divided into two ratios, viz., 30:31 and 31:32. Didymus, however, like Pythagoras, adopted two disjunct tetrachords, and this gave him the perfect fifth in all his scales.

DIDYMUS' CHROMATIC GENERA SCALE.

	C	D♭	D♯	F	G	A♭	A♯	C
Intervening ratios..	15 : 16	24 : 25	5 : 6		15 : 16	24 : 25	5 : 6	
Intervening cents ..	112	70	316	90	112	70	316	60
Length of string ..	120	112·30	108	90	80	75	72	60
Ratios from F to C	3 : 4	4 : 5	5 : 6	1	—	—	—	—
Ratios from C to G	—	—	—	—	3 : 4	4 : 5	5 : 6	1
Ratios ftom C	1	15 : 16	9 : 10	3 : 4	2 : 3	5 : 8	3 : 5	1 : 2
Cents from C	0	112	182	498	702	814	884	120

Here again we have the old Greek interval, ratio 5 : 6.

DIDYMUS' DIATONIC GENERA SCALE.

	C	D♭	E♭	F	G	A♭	B♭	c
Intervening ratio ..	15 : 16	9 : 10	8 : 9		15 : 16	9 : 10	8 : 9	
Intervening cents ..	112	182	204	90	112	182	204	60
Length of string ..	120	112·30	101·15	90	80	75	17·30	60
Ratios from F to C..	3 : 4	4 : 5	8 : 9	1	—	—	—	—
Ratios from c to G ..	--	—	—	—	3 : 4	4 : 5	8 : 9	1
Ratios from C	1	15 : 16	27 : 32	3 : 4	2 : 3	5 : 8	9 : 16	1 : 2
Cents from C	0	112	294	498	702	814	996	1200

Didymus' divisions of the above tetrachord, it will be seen, illustrate perfectly those of the old Grecian diatonic tetrachord. Thus our analysis shows that Didymus was the most conservative of all the Greek musicians who succeeded Pythagoras, and that he refused to be influenced by that philosopher, except in respect to the perfect fifth, which he introduced into his scale by setting up two disjunct tetrachords instead of two conjoint tetrachords, as was the custom of the pre-Pythagorean Greeks.

Yet some commentators have affirmed that Didymus was the most progressive of all the Greek musicians, and that his diatonic scale anticipated our *modern just scale.* This is a grievous mistake and has, we presume, arisen from the circumstances that Ptolemy's diagrams have been read in a reverse way to that intended.

Those so read, certainly make Didymus' diatonic scale somewhat like our major diatonic scale, but this should be regarded as merely a curious coincidence. The question which is really the proper way of reading Ptolemy's diagram can easily be decided by presenting a fac-simile of it.

FAC-SIMILE OF DIDYMUS' DIVISIONS OF THE OCTAVE
IN THE DIATONIC GENERA.*

60, 0	1
67, 30	2
75, 0	3
80, 0	4
90, 0	5
101, 15	6
112, 30	7
120, 0	8

In noticing the above fac-simile of Ptolemy's
diagram of Didymus' diatonic scale it should be
clearly understood that the figures which range
from 60 to 120 do not allude to vibrations but to
lengths of strings. The old Greek musicians never
reckoned by vibrations, no more did Ptolemy
in his diagrams. His method was to indicate
relationship of sounds by comparative lengths of
strings, and the order of the ratios, in Didymus'
diatonic scale, viz., 8 : 9, 9 : 10, 15 : 16, taken into
connection with the order of the figures 60, 67·30,
75, 80, show that the highest note in the
tetrachord is indicated by 60, and the gravest by 80.

* Claudii Ptolemaei. Harmonicorum. Johannes Wallis Editit.

The diagram read upon this authoritative plan shows

DIDYMUS' DIATONIC SCALE.

	$C^{15:16}_{112}$	$D^{\flat 9:10}_{182}$	$E^{\flat 8:9}_{204}$	$F^{8:9}_{204}$	$G^{15:16}_{112}$	$A^{\flat 9:10}_{182}$	$B^{\flat 8:9}_{204}$	C
Intervening ratios .. Intervening cents ..								
Length of string	120	112·30	101·15	100	80	75	67·30	60
Ratios from F to C ..	3 : 4	13 : 15	8 : 9	1	—	—	—	—
Ratios from c to G ..	—	—	—	—	3 : 4	13 : 15	8 : 9	1
Ratios from C	1	15 : 16	27 : 32	3 : 4	2 : 3	5 : 8	9 : 16	1 : 2
Cents from C........	0	112	294	498	702	814	996	1200

and not as follows :—

	$C^{8:9}_{204}$	$D^{9:10}_{182}$	$E^{15:16}_{112}$	$F^{8:9}_{204}$	$G^{8:9}_{204}$	$A^{9:10}_{182}$	$B^{15:16}_{112}$	C
Intervening ratios...... Intervening cents......								
Length of string	—	—	—	—	—	—	—	—
Ratios from F to C	3 : 4	27 : 32	15 : 16	1	—	—	—	—
Ratios from c to G	—	—	—	—	3 : 4	27 : 32	15 : 16	1
Ratios from C..........	1	8 : 9	4 : 5	3 : 4	2 : 3	16 : 27	8 : 15	1 : 2
Cents from C	0	204	386	498	702	900	1088	1200

which *is very nearly our just scale.*

It is satisfactory to find that Professor Helmholtz quotes Didymus' diatonic divisions of the tetrachord as illustrated by—

$$B^{15:16}_{112}\ \Big|\ C^{9:10}_{182}\ \Big|\ D^{8:9}_{204}\ \Big|\ E$$
$$3:4 \qquad\quad 4:5 \qquad\quad 8:9 \qquad\quad 1$$

which is the same as if he quoted as we do.

	C	D^2	$E^{\flat}$	F
	3 : 4	4 : 5	8 : 9	1
Ratios from C	1	15 : 16	27 : 32	3 : 4
Cents	0	112	294	498

In closing this analysis we may venture to express a hope that we have completely proved the fallacy

I

of the claims set up for the Didymus' diatonic scale as approaching to our *modern just scale.*

How far Didymus succeeded in illustrating simple ratios in setting up his scale can be judged by the following table:—

Didymus' ratios from the prime in his

ENHARMONIC SCALE.

1.   31 : 32, 15 : 16, 3 : 4, 2 : 3, 31 : 48, 5 : 8,   1 : 2

IN HIS CHROMATIC SCALE.

1.   15 : 16,   9 : 10, 3 : 4, 2 : 3,   5 : 8, 3 : 5,   1 : 2

IN HIS DIATONIC SCALE.

1.   15 : 16, 27 : 32, 3 : 4, 2 : 3,   5 : 8, 9 : 16, 1 : 2

Chapter XI.

PTOLEMY'S MUSICAL SYSTEM.

AMONGST the vast collection of astronomical, geographical, and other scientific works produced by Ptolemy, the most celebrated voluminous writer of his period (viz., the first half of the second century of our era), we find some remarkable and erudite treatises on music. From one of these works in especial (frequently referred to in these pages), a vast amount of information can be derived concerning the author's own and his predecessors' methods of formulating musical scales. From the excessive devotion which this learned Alexandrian manifested for the study of astronomy, and especially for that abstruse branch of the science which expanded into astrology, it is more than probable that he favoured the Pythagorean views of the correspondence between musical tones and the cosmic order of the universe.

Be this as it may, Ptolemy's profound studies and observations on the experiments of his musical predecessors resulted in a scheme for reducing the number of initial notes in the scale from 12 to 7.

This led him to formulate afresh the divisions of the seven modal scales, to make them commence on a different note of Euclid's canon to those with which they had hitherto been associated, and then to set up his adopted genera scales on the seven initial sounds of the modal scales.

He shows what initial notes he selects, by a diagram, in which upon *Mese*, or the octave to Proslambanomenos, he places the initial of the Dorian scale, which if we regard Proslambanomenos as A was *a*. Upon *Paramese*, which is a whole tone above *Mese*, he places the Phrygian initial, which is then B. Upon *Trite Diezeugmenon*, which is half a tone above *Paramese*, he places the Lydian initial, which is then C. Upon *Paranete Diezuegmenon*, which is a whole tone above *Trite Diezeugmenon*, he places the Mixo-Lydian initial, which is D. Upon *Lichanos Meson*, which is a whole tone below *Mese*, or *a*, he places the Hypo-Lydian initial, which is G. Upon *Parapate Meson*, which is a whole tone below *Lichanos Meson*, he places the Hypo-Phrygian initial, which is F. Upon *Hypate Meson*, which is half a tone below *Parapate Meson*, he places the Hypo-Dorian, which is E.

Thus his initial notes are from lowest to highest

E   F   G   │A│   B   C   D
        │centre│

The Dorian, Ptolemy regards as the centre of his series, and so places three initial scale notes above and three below it.

Having established the relationships of the initial notes of his seven scales, he formulates his first on a the octave to Proslambanomenos and his *Dorian* initial, with the following intervening intervals, as shown by the figures :—

$$1\tfrac{1}{8} \quad 1\tfrac{1}{20} \quad 1\tfrac{1}{8} \quad 1\tfrac{1}{7} \quad 1\tfrac{1}{20} \quad 1\tfrac{1}{9} \quad 1\tfrac{1}{7} \quad 1\tfrac{1}{8} \quad 1\tfrac{1}{20} \quad 1\tfrac{1}{9} \quad 1\tfrac{1}{7} \quad 1\tfrac{1}{20} \quad 1\tfrac{1}{9} \quad 1\tfrac{1}{7}$$

so that Ptolemy's Dorian scale, with A as its initial, was as follows :—

Ptolemy's figs.. Inter. ratios ...	$A^{1\frac{1}{8}}_{8\,:\,9}$	$B^{1\frac{1}{20}}_{20\,:\,21}$	$C^{1\frac{1}{9}}_{9\,:\,10}$	$D^{1\frac{1}{7}}_{7\,:\,8}$	$E^{1\frac{1}{20}}_{20\,:\,21}$	$F^{1\frac{1}{9}}_{9\,:\,10}$	$G^{1\frac{1}{7}}_{7\,:\,8}$
Ratios from A..	1	8 : 9	160 : 189	16 : 21	2 : 3	40 : 63	4 : 7
Cents from A...	0	204	289	471	702	787	969

(SCALE CONTINUED.)

$a^{1\frac{1}{8}}_{8\,:\,9}$	$b^{1\frac{1}{20}}_{20\,:\,21}$	$c^{1\frac{1}{9}}_{9\,:\,10}$	$d^{1\frac{1}{7}}_{7\,:\,8}$	$e^{1\frac{1}{20}}_{20\,:\,21}$	$f^{1\frac{1}{9}}_{9\,:\,10}$	$g^{1\frac{1}{7}}_{7\,:\,8}$	a^{1}
1 : 2	8 : 9	160 : 189	16 : 21	2 : 3	[40 : 63	4 : 7	1 : 2
1200	204	289	471	702	787	969	1200

Ptolemy, with the same order of intervening intervals sets up his other scales on their respective initials. So that he formulates alike all his scales, though he reckons them from their respective initial notes. He gives an extent of two octaves to his scales, but confines them all within Proslambanomenos and Trite Hyperbolœan, *i.e.*, the double octave of Proslambanomenos. He does this by placing below the initial note all the notes necessary to complete his scales which extend above a^1.

Thus, starting his Mixo-Lydian on D, he cannot exhibit the notes beyond a^1, so exhibits the others below D. Thus Mixo-Lydian, $d\ e\ f\ g\ a^1$, then below ;

A B C D E F G A B C

(Read these ten notes from right to left).

This plan has given rise to the statement that Ptolemy reckoned all his modal scales from Proslambanomenos, but that was not the case : he reckoned from the initial notes of each scale. In explanation of his plan, we will suppose we had a pianoforte limited in compass to two octaves, from A to a^1, and we desired to practice the scale of D major, making use of all the notes at command. We should first formulate in our mind the scale of D major, and finger according to that scale, but commence on A, and play A B C♯ before we touch our prime D, and then proceed upwards until we arrive at a^1. We should have played fifteen notes, but should not say we reckoned our scale from A, but from D, the prime of the scale of D.

The following diagram shows all Ptolemy's modal scales :—

PTOLEMY'S MODAL SCALES.

THE MIXO-LYDIAN.

D $\frac{1¼}{8:9}$ E $\frac{1,b}{20:21}$ F $\frac{1¼}{9:10}$ G $\frac{1¼}{7:8}$ A $\frac{1,b}{20:21}$ B $\frac{1¼}{9:10}$ C $\frac{1¼}{7:8}$ | D

THE LYDIAN.

C $\frac{1¼}{8:9}$ D $\frac{1,b}{20:21}$ E♭ $\frac{1¼}{9:10}$ F $\frac{1¼}{7:8}$ G $\frac{1,b}{20:21}$ A $\frac{1¼}{9:10}$ B♭ $\frac{1¼}{7:8}$ | C

THE PHRYGIAN.

B $\frac{1¼}{8:9}$ C♯ $\frac{1,b}{20:21}$ D $\frac{1¼}{9:10}$ E $\frac{1¼}{7:8}$ F♯ $\frac{1,b}{20:21}$ G $\frac{1¼}{9:10}$ A $\frac{1¼}{7:8}$ | B

THE DORIAN (the centre of System).

	Centre	A $\frac{1¼}{8:9}$	B♭ $\frac{1,b}{20:21}$	B $\frac{1¼}{9:10}$	C $\frac{1¼}{7:8}$	D $\frac{1¼}{20:21}$	D $\frac{1¼}{9:10}$	E $\frac{1¼}{20:21}$	E $\frac{1¼}{9:10}$	F $\frac{1¼}{7:8}$	G
Ptolemy's figures											
Intervening ratios											
Intervening cents		204	85	182	231	182	85	231			
Ratios from A		1	8:9	160:189	2:3	40:63	16:21	4:7	1:2		
Cents from A		0	204	289	471	702	787	969	1200		

THE HYPO-LYDIAN.

G $\frac{1¼}{8:9}$ A $\frac{1,b}{20:21}$ B♭ $\frac{1¼}{9:10}$ C $\frac{1¼}{7:8}$ D $\frac{1,b}{20:21}$ E♭ $\frac{1¼}{9:10}$ F $\frac{1¼}{7:8}$ | G

THE HYPO-PHRYGIAN.

F $\frac{1¼}{8:9}$ G $\frac{1,b}{20:21}$ A $\frac{1¼}{9:10}$ B♭ $\frac{1¼}{7:8}$ C $\frac{1,b}{20:21}$ D $\frac{1¼}{9:10}$ E♭ $\frac{1¼}{7:8}$ | F

THE HYPO-DORIAN.

E $\frac{1¼}{8:9}$ F♯ $\frac{1,b}{20:21}$ G $\frac{1¼}{9:10}$ A $\frac{1¼}{7:8}$ B $\frac{1,b}{20:21}$ C $\frac{1¼}{9:10}$ D $\frac{1¼}{7:8}$ | E

Our lettering and the position we have assigned to Ptolemy's modal scales agree with Dr. Wallis's* illustrations, except in one instance.

Making A represent Proslambanomenos the prime, Wallis gives the initial of the Hypo-Phrygian as F♯. We make it F♮, as Ptolemy shows that the place of Hypo-Phrygian was on *Parapate Meson*, and *Parapate Meson* of the Canon is the minor sixth and not the major sixth of Proslambanomenos. In placing three sharps as the signature of the scale on F♯, Wallis makes the scale F♯ G♯ A B C♯ D E F♯. With four flats to the signature the scale would be F G A♭ B♭ C D♭ E♭ which is our reading of Ptolemy's Hypo-Phrygian scale.

Dr. Burney accuses Dr. Wallis† of having mistaken the position of the *Lydian* and *Hypo-Lydian* initials, and says he ought to have placed them half a tone higher, viz., on G♯ and C♯. Dr. Burney grounds this accusation on the statements of Bacchius, senior, who, in his *Introductio Artis Musica*, asks and answers questions. Bacchius‡ asks—

What is the highest note?	...Answer—	The Mixo-Lydian.
What is the next below it?...	„	The Lydian.
How much lower?	„	A half-tone (Hemitonum).
What is next below the Lydian?	„	The Phrygian.
How much lower?	„	A whole tone.

---

* Johannes Wallis, S.S. Th. D., Geometrical Professor Savilianas, Oxoniœ A.D. 1682. Edidit Ptolemæi Harmonicorum.

† General History of Music. C. Burney, Mus. D., F.R.S., A.D. 1786.

‡ Introductio Artis Musica Bacchius, senioris (Meibomius restituit).

Of course, if Bacchius, senior, were explaining Ptolemy's system, his answers would justify Dr. Burney's criticism, but Bacchius, senior, is explaining a system which agrees with the Pythagorean and Euclidean, and according to that system the Mixo-Lydian is B when Proslambanomenos is A, the Lydian is C, the Phrygian D, and the Dorian E, consequently the interval between the Mixo-Lydian and the Lydian is half a tone, the Phrygian is a whole tone lower on the finger board than the Lydian—that is, a whole tone higher in pitch. Thus there is no inference from what Bacchius says which justifies Burney's criticism.

Dr. Burney confesses his inability to explain the Greek systems, as Greek authors contradict each other. It will be found, however, that the Greek authors, when speaking of like systems, agree, and it is only when their remarks on one system are applied (as Burney does) to another, that there is any contradiction in their explanations. It cannot be too carefully borne in mind that there were three principal systems, the *Old Greek*, the *Pythagorean* and the *Ptolemean*, and that under the first two, *hypo* means *higher in pitch* and under the *Ptolemean system below in pitch*.

The most interesting consequence of our analysis of the Ptolemean scales is the proof it affords that their *order* is the origin of that of the ecclesiastical scales, which, like the Ptolemean, is the *Dorian*, the

Phrygian, the *Lydian*, and the *Mixo-Lydian*, with *hypo* scales below in pitch the titled scales.

The Ptolemean modification of the early Greek scales have hitherto been regarded as a curious antique rather than as an authoritative page of musical science. We shall presently show that besides being the origin of the order observed in the ecclesiastical scales there is good ground for believing the *divisions* of four, as well as their titles and order, were adopted by Bishop Ambrose.

PTOLEMY'S GENERA SCALES.

Upon the initial notes of his seven modal scales Ptolemy set up his genera scales.

An explanation of the elaborate arrangement devised by him for so doing would be foreign to our purpose, which is merely to analyse scales, to discover how far the principle of simple ratios was acted upon in formulating them.

It suffices, therefore, to give the results of Ptolemy's division, for the genera scales, and these we learn from his diagrams, which show that some of his scales, though claimed as original were well known under other titles. Others, however, he must be credited with having originated.

Ptolemy's eight genera scales comprise one in the enharmonic genera, two in the chromatic, and five in the diatonic.

PTOLEMY'S ENHARMONIC GENERA SCALE.

	C	W	D♭	F	G	X	A♭	c
Intervening ratios..	45:46	23:24	4:5		45:46	23:24	4:5	
Intervening cents ..	39	73	386		39	73	386	
Length of string ..	120	117·23	112·30	90	80	78·16	75	60
Ratios from F to C..	3:4	23:30	4:5	1	—	—	—	—
Ratios from c to G ..	—	—	—	—	3:4	23:30	4:5	1
Ratios from C	1	45:46	15:16	3:4	2:3	15:23	5:8	1:2
Cents from C........	—	39	112	498	702	741	814	1200

PTOLEMY'S CHROMATIC GENERA SCALES

Entitled "Nostri Mollis Chromatica" and "Nostri Intensi Chromatica."

THE MOLLIS CHROMATICA SCALE.

	C	D♭	D♮	F	G	A♮	A♯	c
Intervening ratios	27:28	14:15	5:6		27:28	14:15	5:6	
Intervening cents.	63	119	316		63	119	316	
Length of string ..	120	115·43	108	90	80	77·9	72	60
Ratios from F to C	3:4	7:9	5:6	1	—	—	—	—
Ratios from c to G	—	—	—	—	3:4	7:9	5:6	1
Ratios from C	1	27:28	9:10	3:4	2:3	9:14	3:5	1:2
Cents from C......	0	63	182	498	702	765	884	1200

THE INTENSI CHROMATICA SCALE.

	C	W	D	F	G	X	A	c
Intervening ratios ..	21:22	11:12	6:7		21:22	11:12	6:7	
Intervening cents ..	80	151	267		80	151	267	
Length of string	120	114·33	105	90	80	76·22	70	60
Ratios from F to C ..	3:4	11:14	6:7	1	—	—	—	—
Ratios from c to G ..	—	—	—	—	3:4	11:14	6:7	1
Ratios from C	1	21:22	7:8	3:4	2:3	7:11	7:12	1:2
Cents from C........	0	80	231	498	702	782	933	1200

The peculiarity of this Intensi chromatica scale is its presenting the ratio 7 : 8 for C-D and 6 : 7 for D-F, just the two ratios required to complete the series of 1 : 2, 2 : 3, 3 : 4, 4 : 5, 5 : 6, 6 : 7, 7 : 8, 8 : 9.

As these divisions of Ptolemy's Intensi chromatica scale have not been elsewhere presented we may regard them as original.

MOLLIS DIATONICA SCALE.

	C	D	E	F	G	A	B	C
Intervening ratios..	120 : 21	♭9 : 10	♭7 : 8		20 : 21	♭9 : 10	♭7 : 8	
Intervening cents ..	85	182	231		85	182	231	
Length of string	120	114·17	102·51	90	80	76·11	68·34	60
Ratios from F to C..	3 : 4	63 : 80	7 : 8	1	—	—	—	
Ratios from c to G ..	—	—	—	—	3 : 4	63 : 80	7 : 8	1
Ratios from C......	1	20 : 21	6 : 7	3 : 4	2 : 3	40 : 63	4 : 7	1 : 2
Cents from C	0	85	267	498	702	787	969	1200

In the above divisions we again meet with the ratios 6 : 7 and 7 : 8, viz., 6 : 7, C-E♭, cents 267, and E♭-F, 231 cents.

TONICI DIATONICA SCALE.

	C	D	E	F	G	A	B	C
Intervening ratios..	27 : 28	♭7 : 8	♭8 : 9		27 : 28	♭7 : 8	♭8 : 9	
Intervening cents..	63	231	204		63	231	204	
Length of string	120	115·43	101·15	90	80	77·9	67·30	60
Ratios from F to C..	3 : 4	7 : 9	8 : 9	1	—	—	—	
Ratios from c to G ..	—	—	—	—	3 : 4	7 : 9	8 : 9	1
Ratios from C......	1	27 : 28	27 : 32	3 : 4	2 : 3	9 : 14	9 : 16	1 : 2
Cents from C	0	63	294	498	702	765	996	1200

Ptolemy may be credited with originality in some of the above divisions, as we have not elsewhere discovered the use of the ratios.

DIATONICI DIATONICA SCALE.

	C	D	E	F	G	A	B	C
Intervening ratios....	243 : 256	♮8 : 9	♮8 : 9		243 : 256	♮8 : 9	♮8 : 9	
Intervening cents....	90	204	204		90	204	204	
Length of string	120	113·54	101·15	90	80	75·56	67·30	60
Ratios from F to C ..	3 : 4	64 : 81	8 : 9	1	—	—	—	
Ratios from c to G ..	—	—	—	—	3 : 4	64 : 81	8 : 9	1
Ratios from C........	1	243 : 256	27 : 32	3 : 4	2 : 3	81 : 128	9 : 16	1 : 2
Cents from C	0	90	294	498	702	792	996	1200

The above is similar to the old Dorian tetrachordal scale, also to the Pythagorean and Euclidean diatonic.

INTENSI DIATONICA SCALE.

	C	D	E	F	G	A	B	C
Intervening ratios....	15:16	♭8:9	♭9:10		15:16	♭8:9	♭9:10	
Intervening cents....	112	204	182		112	204	182	
Length of string	120	112·30	100	90	80	75	66·40	60
Ratios from F to C ..	3:4	4:5	9:10	—				
Ratios from c to G ..	—	—	—	—	3:4	4:5	9:10	1
Ratios from C........	1	15:16	5:6	3:4	2:3	5:8	5:9	1:2
Cents from C	0	112	316	498	702	814	1018	1200

The above is the old Grecian diatonic set up with disjunct instead of conjoint tetrachords.

EQUABILIS DIATONICA SCALE.

	C	D	E	F	G	A	B	C
Intervening ratios..	11:12	♭10:11	♭9:10		11:12	♭10:11	♭9:10	
Intervening cents..	151	165	182		151	165	182	
Length of string ..	120	110	100	90	80	73·20	66·40	60
Ratios from F to C	3:4	9:11	9:10	1				
Ratios from c to G	—	—	—	—	3:4	9:11	9:10	1
Ratios from C......	1	11:12	5:6	3:4	2:3	11:18	5:9	1:2
Cents from C	0	151	316	498	702	853	1018	1200

In the above scale we have illustrations of the use of ratios 10 : 11 and 11 : 12, which, on reference to the harmonic scale, chapter XV. will be seen to be harmonic intervals, but not likely to have been known as such to Ptolemy.

CHAPTER XII

THE ECCLESIASTICAL SCALES.

WHETHER the early Christian hymns, chants, and simple services were revivals of Hebrew fragments transmitted from Apostolic days, or were taken directly, as some assert, from the Greeks, is a mooted question, by no means authoritatively decided, but it is quite certain that at the period between the third and fourth centuries, when Greek art was admitted to be the ruling influence in all matters of literature and science, especially after the Council of Nice, the music used in the Christian churches was arranged according to the Greek modes, for history records that about the year A.D. 388, Ambrose, Bishop of Milan, appointed four of the most suitable of the Greek modes to be used in the services of the Christian Church, and the four were those then known under the titles

THE DORIAN,
THE PHRYGIAN,
THE LYDIAN,
THE MIXO-LYDIAN.

The initial notes are usually given as D, E, F, G. It must be borne in mind that the prefixes to these scales, though borrowed from the Greeks, must

be distinguished as the *Ecclesiastical Dorian,*
Ecclesiastical Phrygian, &c., the scales implied being
different from the Old Greek and Pythagorean.

The above four modes have ever since been associated with the name of Bishop Ambrose, and called
the Ambrosian scales. They belong to a series which
became entitled *authentic, i.e.,* superior or correct.
The term, however, grew to be applied to chants and
hymns having their final on the initial note.
History also recounts that following the example
of Bishop Ambrose, Pope Gregory, A.D. 598-604,
selected three more modes for the use of the Christian
Church.   These were known at the time as

THE HYPO-DORIAN.
THE HYPO-PHRYGIAN.
THE HYPO-LYDIAN.*

These scales were called "plagal" scales, *i.e., inferior,*
though the term was used to indicate that the final
note must be a fourth above the initial note.   Pope
Gregory also made the Dorian *plagal* besides being
authentic. As plagal it was designated

THE HYPO-MIXO-LYDIAN.

In process of time, as the number and power of the
ecclesiastical party increased, the desire to vary the
music of the Church gave rise to the adoption of

* The initial notes are usually given as A B C—thus making the series of
initial notes A B C D E F G.

other modes, but as innovations were stoutly resisted by the more conservative ecclesiastics, disputes arose on this point, until it was ultimately agreed to refer the matter to the illustrious and pious Karl der Grosse, who decided that the introduction of four more ecclesiastical scales might promote rather than retard the best interests of ecclesiasticism.* Armed with this powerful authority, the innovators added to the scales already in use the following :—

The Eolian, with its initial on the 5th note of the Dorian scale, and the Ionian, with its initial on the 7th note.

To the Eolian was given a hypo on the 4th below, and to the Ionian a hypo on the 4th below.

These additions could scarcely be regarded as of much importance, the *Eolian* being only an octave above the Hypo-Dorian. It was allowed to be *authentic*, however, whilst the Hypo-Dorian was *plagal*. *The Ionian* was the scale an octave above the Hypo-Lydian, but was authentic, whilst the Hypo-Lydian was plagal.

The Hypo-Eolian was identical with the Phrygian, but was plagal, whilst the Hypo-Ionian was identical with the Mixo-Lydian, but was plagal.

---

* Six modes or scales were submitted to Karl der Grosse as desirable additions, but the resolute monarch concluded only to permit four to be used ; the two others were rejected.

It should be observed that Bishop Ambrose placed the Phrygian above the Dorian, the Lydian above the Phrygian, the Mixo-Lydian above the Lydian; and that the initial notes of his four scales are represented by D, E, F, G, but any four letters would do, implying that the tones they represent are separated by a tone, a half tone, and a tone. The Ambrosian-Dorian scale, according to the lettering, is,

$$\overgroup{\text{D E F G}} \quad \overgroup{\text{A B C D}}$$

and the Phrygian

$$\overgroup{\text{E F G A}} \quad \overgroup{\text{B C D E}}$$

whilst the Pythagorean Dorian, with D as its initial, should be

$$\overgroup{\text{D E♭ F G}} \quad \overgroup{\text{A B♭ C D}}$$

and the Pythagorean Phrygian should be, with E as its initial,

$$\overgroup{\text{E F♯ G A}} \quad \overgroup{\text{B C♯ D E}}$$

The marked difference between the Pythagorean and the Ecclesiastical scales of like titles, as illus-trated above, has led to, though not justified, the statement that the Ecclesiastical scales are distinguished by the names of their Greek prototypes,

K

though not really identical with them. The state-ment and the belief, however, are equally erroneous.

Bishop Ambrose and Pope Gregory adopted the titles and order of the scales in vogue in their times amongst the Greeks—that is to say, the Old Greek titles, but the Ptolemean order of them, which were, reading from the highest to the lowest,

THE MIXO-LYDIAN.
THE LYDIAN.
THE PHRYGIAN.
THE DORIAN.
THE HYPO-LYDIAN.
THE HYPO-PHRYGIAN.
THE HYPO-DORIAN.

The mistake of stating and believing that the Ecclesiastical scales are not identical with the Grecian, appears to have arisen from the idea that *Greek scales* with like titles were identical with each other—that the Dorian of one period was the Dorian of another—and that there is no difference between the scales of one era and another, although different philosophers undertook to modify, alter, or amend them.

The only points really open to discussion are, did Ambrose adopt Ptolemy's various whole tones of $8:9$, $9:10$, and $7:8$, and Ptolemy's half tones of $20:21$, or did he make all his whole tones $8:9$, and his half tones $243:256$, like the Pythagoreans?

It is hardly likely that in adopting Ptolemy's order of scales and his succession of tones and half tones that Bishop Ambrose refrained from adopting his exact intonation. The proof is wanting that he used the Ptolemean ratios 20 : 21 between the second and third notes of his Dorian scale, and the ratio of 9 : 10 between the third and fourth notes of that scale, but the inference is strong that he did, if as it has been said, the sixth in the scale in the old Ambrosian missals is minor. This only happens in the Pythagorean scales when the second is a whole tone, in the case of the Locrian series. As there is cause for believing that Bishop Ambrose adopted one Ptolemean peculiarity of intonation, it may reasonably be concluded he adopted more than one. The tradition is, that Ambrose's scales differed from Pope Gregory's, in which the whole tone is ratio 8 : 9, the half tone is ratio 243 : 256, and the sixth of the Dorian is major, as will presently be shown was the case. If the Ambrosian series differed only in the sixth being minor instead of major, it would have been most likely so said, but differing, as we believe they did, in many respects, the inference is that they differed in more than could concisely be explained.

Bishop Ambrose's Dorian scale, with Ptolemean intonation would be as follows, and we are inclined to believe that it was so :—

	D	E	F	G	A	B♭	C	D
Intervening ratios ..	8 : 9	20 : 21	19 : 10	7 : 8	20 : 21	9 : 10	7 : 8	
Intervening cents...	204	85	182	231	85	182	231	
Ratios from D	1	8 : 9	—	—	2 : 3	—	4 : 7	1 : 2
Cents from D	0	204	289	471	702	787	969	1200

It is recorded that Pope Gregory not only decided to appoint four more Grecian modes to the service of the Roman churches—he really only appointed three, for the one he called the Hypo-Mixo-Lydian was the Dorian made plagal—but that he reformed the music then in use. We infer by this that he altered the intonation or tuning of the Ambrosian scales.

At all events, the tuning of the Ambrosian scales, as handed down to us *since* Pope Gregory's time, has been regarded as Pythagorean—that is, the whole tones are ratio 8 : 9, the half-tone ratio 243 : 256, cents 204 and 90 ; so that instead of the Ptolemean tuning, which, as expressed by cents, from the initial note E, was

	E	F♯	G	A	B	C	D	E
	E_{204}	$F^{\sharp}_{85}$	G_{182}	A_{231}	B_{85}	C_{182}	D_{231}	E
Cents	0	204	289	471	702	787	969	1200

Gregory gave the following tuning to that scale :—

	E	F	G	A	B	C	D	E
	E_{90}	F_{204}	G_{204}	A_{204}	B_{90}	C_{204}	D_{204}	E
Cents	0	90	294	498	702	792	996	1200

thus effecting an alteration in the Ptolemean tuning, whilst preserving the Ptolemean order of the modal scales.

The initial notes of the Gregorian Hypo (plagal) scales were perfect fourths below the respective authentic, but according to the letters handed down to us, all the *notes* of the plagal scales were not perfect fourths below those of the authentic.

Only seven differently-constructed scales were associated with the rites of the Romish Church. Six of these formed the series called Authentic. The others were similarly constructed, but were called Plagal scales, and the use of them was subject to an ecclesiastical law, which does not fall within this treatise to enlarge upon. Regarding ecclesiastical scales since the time of Pope Gregory, as formulated with the Pythagorean ratios of 8 : 9 and 243 : 256, cents 204 and 90, it is evident that the beautiful in sound, dependent as it is, and ever must be, on the use of intervals determined by the simplest ratios, was no more evolved by the ecclesiastics than by the Greek philosophers.

Admitting harmony (*i.e.*, combined sounds) to be one of the most potential of all elements in the production of delightful music, what combinations of an endurable character could be formed amongst tones, the only established consonances of which were the perfect fourth and perfect fifth? In both the Greek and Ecclesiastical scales, the third and sixth having to be expressed by complex ratios, no consonance of tone could be formed with them, and this point alone would be sufficient to show the identity of the Greek and Ecclesiastical scales.

Of the Greek philosophers it might be said, as of the mediæval alchemists, they hovered on the borders of the grand discovery they sought to compass again and again, but when the moment arrived for the triumphant culmination of all their labours, the result only proved that they had yet to sound the key-note which could formulate the just and perfect scale. The ecclesiastics were only imitators and not originators, and thus their scales fail to prove any advance upon those of the Pythagorean or Ptolemaic periods.

RETROSPECT OF THE ANALYSIS OF THE ECCLESIASTICAL SCALES.

Our investigations and analyses of the Ecclesiastical scales bring to light the fact, that Ambrose, a learned advocate (prior to his assumption of the bishop's mitre), and a man who could not fail to be familiar with Greek art and literature, in selecting his four modes, adopted the Ptolemean order of them; so that when three centuries after his time Pope Gregory and the ecclesiastics of the Christian Church desired to enlarge upon and add variety to their musical services, they had but to adopt the titles of the additional three scales formulated by Ptolemy and add a Hypo to the Mixo Lydian, to establish the eight scales they required. The four other scales subsequently devoted to the service of the Christian Church were similar, or an octave higher, to four others. The use made of them accounts sufficiently for the ecclesiastical scales being reckoned as twelve in number. The idea that the Ecclesiastical and Greek scales were only identical in name is, as we have shown, a groundless assumption.

RETROSPECT OF THE FOREGOING ANALYSES OF ALL THE GRECIAN SCALES.

It needs only a careful study of the modes and scales laid down in the preceding pages to discover that the aim of the early musicians, not only of the classic Grecian ages, but also of still earlier periods, and amongst less cultivated peoples than the Greeks, was to formulate a series of sounds whose relationship to a given note could be described with mathematical precision, as well as show agreement with the results of the known laws of the natural sciences. The Greeks, who unquestionably devoted far more time, skill, and philosophic experiment to this subject than other nations, seem often to have been on the eve of achieving the great discovery and finding the link which connected the beautiful in sound with the fundamental principles of just mathematical law.

The reiterated queries of Pythagoras and Aristotle,

concerning consonances, the bold revolt of Aristoxenes from the severe rule of the mathematicians,
and his attempt to refer the arbitration of the
beautiful in sound principally to the ear, all mark the
earnest research with which the final solution of the
problem was pursued.   There seem to have been
occasional rifts in the clouds which darkened the
mental horizon, when the sunlight of a true union
between science and æsthetics might have been perceived and reduced to a steady and permanent
illumination.   For example, the great underlying
principle of determining intervals by simple ratios
gave birth to the celebrated Pythagorean mode of
obtaining the notes of the scale by perfect fifths.

Why the experimenters failed has already been
shown, and yet those same experimenters stood on
the very verge of achievement, and only failed when
they misapplied their own beautiful theory.   Again,
it is evident that the twelve notes of the chromatic
scale were known to the Greeks, for upon twelve
notes within the octave they set up their Genera
scales, using the twelve sounds as initial notes.

In commenting on the subject, Mr. Ellis, in a note
to his translation of Helmholtz, says, " It is by no
means an unimportant fact that a flute was found in
the royal tombs at Thebes, in Egypt (now in the
Florentine museum), with an almost perfect scale of

semitones for about an octave and a half.   .   .   .
And representations of such flutes are found in the
oldest Egyptian monuments." So that the Greeks
must have been well aware that other nations
made a more extended use of chromatic intervals
than they did.

The result of our researches and analyses fully
shows that whilst, to discover the beautiful in
sound and the just in science, were the aim and
object of a host of philosophers, great nations rose,
flourished, and sank into decay without the mystic
face of success being unveiled to the gaze of the
antique sage, who only saw dimly the goal of his
great ambition.

Whether the early Christian ecclesiastics were too
deeply imbued with reverence for Greek authority to
attempt any marked innovations on their systems of
musical order, or whether they themselves lacked
the clue which should guide them through the mazes
of tonal mysteries, it would be now impossible to
say.   Certain it is that they only repeated classic
errors without developing any fresh modes of cor-
recting them.

How this great but long and vainly-sought-for
desideratum was at length attained we will now
endeavour to trace out.

Chapter XIII.

THE JUST SCALE AND ITS DISCOVERY.

DISCOVERIES are very frequently effected long after their pursuit has been abandoned, and what appears to be fortuitous accident often discloses the secret which has eluded the most careful investigation.

So has it been with the discovery of that just scale on which the beautiful in sound depends.

Long after ancient Greece had fallen into decay, and Christian ecclesiastical musicians, ever aiming to add the charm of music to their services, had adopted Greek errors along with Greek art—in fact, as late as the tenth century—the musicians of Europe, in attempting to combine tones into parts, hit upon the system of "descant," and as descant was in reality the first progenitor of harmony, and harmony could only be evolved from consonances, an effort was made to reconstruct the divisions of the ecclesiastical scales and to add to the number of consonances.

More than a thousand years before the discovery of descant, Aristotle had asked "why the consonance of the octave alone was sung." The contemporaries of the great Stagirite were unable to answer him ; but modern science boldly replies that the complex ratios of the intervals of the third and sixth in the Pythagorean system made them unfit to combine with other sounds.

Without reiterating the defects of the ancient scales, it is enough to say that the principle of combining chords in harmonious relations depends quite as fully on intervals expressed by whole numbers or simple ratios, as does a just succession of tones, and that so long as the Western nations contented themselves with the adoption of the early ecclesiastical scales, satisfactory combinations of musical intervals were as impossible as with the Greeks.

Descant was originally the performance of two different tunes simultaneously, the airs being so constructed that the voices should sound, at the same time, fourths, fifths, or octaves.

The Flemish monk Hucbald writes of it as an invention called *Organum*, or Diaphony, and describes it "as two voices singing together fourths and fifths, with occasional doublings of octaves." It was also called *Discantus*, and O. Paul, in his history of the clavier, alluding to *Discantus*, says : "It was no uncommon thing for a solemn liturgical chant to be coupled up in this way with songs of a free, not to say slippery character."

Helmholtz, tracing up the gradual development of Polyphonic music, says: "The principle of descant was fertile, and Polyphonic music was the result. Different voices, each proceeding independently and singing its own melody, had to be united in such a way as to avoid dissonances, or only create such as were readily passed over. Thus it was that music increased in richness as parts multiplied, but the establishment of an artistic connection between the parts in vocal music gave rise to a new invention, and this consisted in causing a musical phrase which had been sung by one voice to be repeated by another." "Thus arose canonical imitation, a device which dates from about the 12th century, and soon obtained much popularity."

It must not be assumed, however, that descant, or canonical imitations, formed any corrective to the imperfections of the Greek ecclesiastical scales.

Beyond the charm of novelty which descant and canonical imitation imparted to the practice of vocal music, their chief use as an element of progress was the impulse they communicated to the practice of combining sounds, or forming polyphonic music, but no true science of musical progression was constructed, nor correct scales evolved, by massing voices together in a mere pastime.

The necessity for forming just scales and

thoroughly related intervals was first felt by those who desired to produce true *Harmony*, or a series of satisfactorily related sounds for each and both voices in combination.

In process of time, and when it became evident that fixed and scientific methods of musical composition must be determined upon, the principles of Greek art and the mathematical proportions laid down by the Greeks were discussed anew. Then it was that the Pythagorean system of admitting into the scales intervals with complex ratios was proved to be wholly impracticable. The results accruing from such discussions in the light of modern science and the continual struggles of the human mind for advancement were inevitable.

The Pythagorean system was found wanting. Complex ratios were deemed inadmissible and impracticable, and the adoption of the fundamental principle of simple ratios, as the basis of pure melody and true harmony, was felt to be the necessity of musical progress.

Thus it will be seen, that so long as music was limited to one part, or what is now called simple melody, the Greek scales, with all their imperfections, both in theory and practice, might have continued to exert an authoritative and conventional sway over the art of music, but from the time when Harmony, even in its incipient stages of development, began to suggest the beauty of combined tones and the necessity of exploring fresh tonal laws, the Greek systems were not only called into question but the cause of their failure became clearly manifest.

Once freed from the sense of proscriptive deference to classical authority, which the ecclesiastical music of the mediæval ages had done much to foster, modern musicians commenced a thorough and searching analysis into the best modes of reconstructing the scale.

They soon found that in order to make the tonic and mediant or third note sound well in combination in a major key, the Pythagorean major third ratio 64 : 81, 408 cents, must be lowered to the simpler ratio 4 : 5, cents 386.

Thus could be constituted a perfectly harmonious chord of the major third and perfect fifth. Moreover, in order to make the superdominant or sixth note consonant with the tonic in a major key, the ratio of 3 : 5 was adopted, 884 cents, instead of ratio 16 : 27, cents 906.

In the same way, to obtain a major third between the fifth and seventh notes of the major scale, the ratio of the seventh note was lowered to 8 : 15, cents

1088, from ratio 128 : 243, cents 1110. The series thus evolved is represented as follows :—

Intervening cents...	C_{204}	D_{182}	E_{112}	F_{204}	G_{182}	A_{204}	B
Ratios from C	—	8 : 9	4 : 5	3 : 4	2 : 3	3 : 5	8 : 15
Cents from C ......:	—	204	386	498	702	884	1088

By lowering the Pythagorean third to 386 cents, and the Pythagorean sixth to 884 cents, all the notes of the scale find consonances amongst each other. By lowering the third and the major seventh of the Pythagorean ecclesiastical scales to ratio 8 : 15, cents 1088, the ratio 128 : 243, cents 1110, was eliminated, and B not only made a just major third to G but a just major sixth to D and a perfect fifth to E. By giving to the interval of the major seventh the ratio 8 : 15, there were also established an augmented fourth between F and B and a diminished fifth between B and F above, which though (not held to be consonant intervals) are not dissonant to the ear.

Thus, in the construction of a true scale the impossibility of combining chords into harmony, a defect which was peculiarly characteristic of the Pythagorean school, was at last wholly removed ; a natural succession, as well as the opportunity for forming the most perfect and harmonious combination of tones, was presented ; the just major diatonic scale was completed ; and the exemplification of the theory that the beautiful in sound depends upon duly related intervals was at last successfully illustrated.

THE JUST MINOR DIATONIC SCALE.

This scale was formed by making the third and sixth of the series minor—*i.e.*, by adopting the ratios 5 : 6 and 5 : 8 instead of 4 : 5 and 3 : 5. A perfect fifth was given to the mediant (the minor third of the key), and that introduced an exchangeable note in the minor series.

One form of the minor scale is—

	C	D	E	F	G	A	B♭	C
	8:9 204	15:16 112	9:10 182	8:9 204	15:16 112	64:75 274	15:16 112	
Ratios from C ..	1	8:9	5:6	3:4	2:3	5:8	8:15	1:2
Cents from C ..	0	204	316	498	702	814	1088	1200

Another form used generally as a descending scale is

	C	D	E♭	F	G	A♭	B♭	C
	:9 204	15:16 112	9:10 182	8:9 204	15:16 112	8:9 204	9:10 182	
Ratios from C	1	8:9	5:6	3:4	2:3	5:8	5:9	1:2
Cents from C	0	204	316	498	702	814	1018	1200

A series for the minor scale has also come into use by adopting a major sixth as well as major seventh in the *ascending* scale. This series avoids the wide gap which would otherwise exist between a minor sixth and major seventh, the leading note or seventh being generally major in the ascending scale. The major sixth is, however, only used in an *ascending* scale.

MINOR SCALE WITH MAJOR SIXTH.

$C^{8:9}_{204}$	$D^{15:16}_{111}$	$E^{\flat 9:10}_{182}$	F	$G^{10:10}_{152}$	$A^{\sharp 8:9}_{204}$	$B^{\natural 15:16}_{112}$	C
1	8:9	5:6	3:4	2:3	3:5	8:15	1:2
0	204	316	408	702	884	1088	1200

The adoption of the ratio 5 : 8 for the sixth in minor scales was required in order to give to that interval a perfect fourth below, to give the sub-dominant of the tonic a minor third above and the octave to the tonic a major third below.

With the improved ratios, as shown above, the practice of combining sounds became general, and to Homophonic was added Polyphonic music.

Chapter XIV.

VIBRATORY ACTION: ITS DISCOVERY AND APPLICATION.

THE exemplification and correction of the old Greek theory concerning the beautiful in sound by a practical adaptation of the principle of simple ratios to the intervals of the diatonic scale became too obviously the only rational solution of tonal mysteries in the light of modern science to admit of denial.

The universal accordance of opinion on this subject throughout the civilized world, has operated almost in effect like a new discovery, but one from which there can be no appeal: hence it has only remained for the musicians of our own time to analyse the relationship which exists between the methods of antiquity (especially in classic periods), to follow out the gradual evolution of modern art to the culminating point of present achievement, and consider what factors have been most instrumental in developing the best methods of our own time. In tracing out the unfoldment and growth of our modern systems, we are necessarily led to inquire what means have been employed to establish the strict grammatical rules upon which our present status of art is founded, the most inspiring and powerful combinations of sounds have been arranged, and mathematical law and æsthetic taste have been reconciled.

Progress in all these directions has been constant and unbroken, resulting in rules which constitute the grammar of harmony and composition, too well established to need comment, but far too diffuse to permit of description in this treatise.

To recur once more to the Greek methods, it must be universally admitted that their experiments were conducted through the agency of stretched strings, the lengths and thicknesses of which, together with their mathematical proportions, determined the result of their experiments.

Whilst such close observers and tireless experimenters as the Greeks must have noticed the fact—apparent to every eye—that vibratory action accompanied every impulse produced upon a stretched string, it is quite certain that they had none of our modern mechanical appliances for gauging vibratory motions. Hence there is no reason to suppose they departed from the principle of determining the pitch of tones by the various lengths of string experimented with.

Whatever may have been the opinions entertained by the Greeks, or indeed by any mere theorists, on this subject antecedent to the period of Galileo, it must be admitted that to him may be attributed a series of observations which culminated in the science of vibratory action.

To obtain a correct standpoint for definitions on this subject it may be stated in brief, that Galileo's experiments with a pendulum proved that all its movements are accomplished in the same periods of time, whether caused by a slight or a powerful impulse of disturbance—in technical phraseology, "whatever be the amplitude of the vibrations."

Proceeding to experiment upon this basic law, physicists soon found :—

1. That tones of music were caused by, or produced upon the air, a set of undulatory movements termed vibrations.

2. That the octave of any musical sound has double the number of vibrations to the prime tone.

3. That the fourth to a prime tone has vibrations as 4 to 3—in fact, that the simple ratio of the length of string existed also for the exact vibrational numbers of the tones produced.

4. That the precise number of vibrations produced in the air belong to the pitch of tones, independent of the means of producing them or the nature of the instrument sounded.

In this discovery, of course, exists the cause of the

just relationship between tones and vibrations, and herein also is the reason why whole numbers and simple ratios strike upon the ear with the effect of pleasure, or produce the beautiful in sound, *i.e.*, why whole numbers and simple ratios should exist between intervals, whether in succession or combination.

Without entering at this point upon the physical causes which underly the agreeable sensations produced upon the ear by whole numbers or simple ratios, or the still more subtle question of the coalescences which do and must exist between numerical proportions in all forms of science, it is enough to say, in the coalescence of vibrational numbers was found the reason why the succession of sounds in the just scale should determine the intervals of the third, fourth, fifth, sixth, and eighth notes. The second and seventh were also in due order recognised as essential to fill up the intervals between the prime and the octave.

The octave being invariably found to register vibrations as 2 to 1, it was discovered that results of the same definite kind existed between a prime and its fourth, giving the ratio as 4 to 3, with the fifth as 3:2, and with the major third as 5:4.

The propriety of adopting the ratio of 8:9 for the second note of the scale was corroborated by observing that such a ratio depicts the relationship existing between the fourth and fifth notes of the scale, besides providing a minor third of 294 cents to the fourth note, a perfect fourth to the fifth note, and a major sixth below the leading note of the scale.

In determining the ratios of these intervals by vibrations, it becomes apparent that the more frequent the coalescence of vibrations the more delightful and harmonious is the combination. Hence, the cause of the beautiful in sound might be clearly traced to the vibratory action of the air, in which the movements could be indicated by simple ratios.

The mistake of attributing to the *lengths* of strings what was due to vibratory action upon the air was most natural and reasonable with the ancients. Had rods fixed at both ends, or closed pipes been substituted for strings and open pipes, the error might have been discovered, the natural divisions of the former yielding results different from the latter. The ancients were able to apprehend clearly why the octave of a note given on a stretched string should be as 1 to 2, but could not have explained why a flame from a gas burner, a piece of metal, or any other resonant body, should sound an octave to a pipe or string. In the above examples of producing sounds the relative lengths of a stretched string are

wanting. In all probability the Greeks were not
much acquainted with the nature of vibratory
action, even in the case of wind instruments or
instruments of percussion. Hence, it is doubtful
if they would have appreciated the utility of
registering vibrations, even, if they had possessed
the means of doing so. The doctrine of simple
ratios for the evolution of the beautiful appears
to have been a veritable inspiration to the old
philosophers, for it could not have resulted from
acoustic observations.

By the discoveries and application of vibratory
calculations one of the most profound theorems of
tonal science and beauty dawns upon the mind,
for, the formation of a just as well as a beautiful
scale of music, once the great tonal problem
of the ancients, resolves itself into a set of fundamental principles, which future researches may
elaborate but can never destroy.

That which constitutes the truly beautiful in
sound, whilst now and ever dependent upon mathematical proportions, is no longer the veiled Isis of
musical science, but discloses itself, like all natural
laws, in the simplest form, as the true relation
which the whole numbers bear to each other and to
the air, by the undulatory motions which different
degrees of pitch occasion. Thus music is the per-

former, and the air the instrument, the latter keeping a faithful and unerring registry, and recording to every capable observer the pitch of every tone that sets its waves in motion.

Chapter XV.

THE HARMONIC SCALE.

THIS scale is called the "scale of nature," or that series of tones which is produced by the *partials** of a vibrating elastic body, such as a stringed or wind instrument, *i.e.*, by the sounds which the elastic body yields through vibratory movements of its aliquot parts in addition to the prime sound or the sound of the complete body.

Whatever may be the number of the vibrations of a string producing the fundamental note, that number multiplied by the figures of the arithmetical progression (2, 3, 4, 5, 6, 7, 8, 9, 10, &c.) gives the number of the respective higher partials. Thus, if the fundamental note or first partial is the result of 30 vibrations, the second partial is the result of 30×2 the next partial is the result of 30×3, the next partial 30×4, and so on to the last appreciable vibratory tone. In the following diagrams the figures 1, 2, 3, &c., show the arithmetical progression, also the ratios of the partials to the prime and to each other.

---

* We adopt Professor Helmholtz's term partials in preference to the one hitherto more commonly used—namely, harmonics—the word partial being more strictly definitive of the subdivisions of a fundamental note than harmonics. In this sense, therefore, the fundamental note is the first partial ; the first harmonic is the second partial or first upper partial.

The letters F, A, C, D, E, E♭ are symbols of the sounds indicated by the ratios, when F is prime. Thus, A is the sound, ratio 4 : 5 to F, *i.e.*, its major third. D is the sound, ratio 16 : 27 to F, and not D 3 : 5 to F. W, X, Y, Z stand for sounds which the letters of the music alphabet fail altogether to indicate.

W stands for a sound 30 vibrations acuter than the sound of the preceding note or scale letter; X stands for a sound 60 vibrations acuter; Y for a sound 90 vibrations acuter; and Z for a sound 120 vibrations acuter.

The third row of figures gives the relative number of vibrations of the partials under which they are placed. The fourth row shows the ratio of the harmonic sounds to the prime, or to one of the octaves of the prime.

THE HARMONIC SCALE.

Ordinal number of partials	1	2	3	4	5	6	7	8
Intervening ratios..	F 1:2	F 2:3	C 3:4	F 4:5	A 5:6	C 6:7	E♭ 7:8	F 8:9
Pitch number	30	60	90	120	150	180	210	240
Ratios from prime or octave to prime	1	1:2	2:3	1:2	4:5	2:3	4:7	1

(HARMONIC SCALE CONTINUED)

Ordinal number of partials	9	10	11	12	13	14	15	16
Intervening ratios..	G 9:10	A 10:11	W 11:12	C 12:13	W 13:14	E♭ 14:15	E♯ 15:16	F
Pitch number	270	300	330	360	390	420	450	480
Ratios from prime or octave to prime	8:9	4:5	8:11	2:3	8:13	4:7	8:15	1:2

(HARMONIC SCALE CONTINUED.)

17	18	19	20	21	22	23	24
W	G	W	A	W	X	Y	C
510	540	570	600	630	660	690	720

(HARMONIC SCALE CONTINUED.)

25	26	27	28	29	30	31	32
W	X	D	E♭	W	E♮	W	F
750	780	810	840	870	900	930	960

(HARMONIC SCALE CONTINUED.)

33	34	35	36	37	38	39	40
W	X	Y	G	W	X	Y	A
990	1020	1050	1080	1110	1140	1170	1200

(HARMONIC SCALE CONTINUED.)

41	42	43	44	45	46	47	48
W	X	Y	Z	B♮	W	X	C
1230	1260	1290	1320	1350	1380	1410	1440

The notes or letters enclosed in ⬜ show the provision made by harmonics for the formation of the just major scale. The ratios indicated by the ordinal number of the partials immediately above these ⬜ brought to their lowest denominations will be seen to be 8 : 9, 9 : 10, 15 : 16, 8 : 9, 9 : 10, 8 : 9, 15 : 16, which are the intervening ratios of the just scale—thus 24 : 27 = 8 : 9, 27 : 30 = 9 : 10, 30 : 32 = 15 : 16, &c.

In the harmonic scale nature has provided for various descriptions of scales. Thus what is termed the just major scale is provided for by adopting the partials 24, 27, 30, 32, 36, 40, 45, 48, of any sounds taken as a prime. Assuming that F is taken as a prime, then (as in the foregoing diagrams) the 24th partial is C, with 720 vibrations. Starting from C, and adopting the 27th partial of F, we have D with 810 vibrations, ratio to C 8 : 9. Adopting the 30th partial of F we have E; ratio to C, 4 : 5. Adopting the 32nd partial to F we have F, 960 vibrations; ratio to C, 3 : 4. Adopting the 36th partial of F we have G; ratio to C, 2 : 3. Adopting the 40th partial, 1,200 vibrations, we have A; ratio to C, 3 : 5. Adopting the 45th partial, with 1,350 vibrations, we have B; ratio to C, 8 ÷ 15; and adopting the 48th partial, with 1,440 vibrations, we have C; ratio to C, 1 : 2.

THE JUST MAJOR SCALE OF C, SET UP FROM PARTIALS OF F.

Ordinal number of partials	24	27	30	32	36	40	45	48
	C	D	E♮	F	G	A	B♮	C
Pitch numbers	720	810	900	960	1080	1200	1350	1440
Ratios from C	1	8 : 9	4 : 5	3 : 4	2 : 3	3 : 5	8:15	1 : 2
Cents from C	0	204	386	498	702	884	1038	1200

THE DIATONIC MINOR SCALE.

The partials do not give the minor scale *as we define it*—namely, with ratios 8 : 9, 5 : 6, 3 : 4, 2 : 3, 5 : 8, 5 : 9, 1 : 2, from prime, consequently we cannot call it a natural scale—that is, a scale pointed out by nature as a *true* scale—but the partials give a minor scale, differing from the one in use in certain respects—that is, a certain kind of minor scale. These partials are 24, 27, 28, 32, 36, 38, 45, 48.

Starting from C, 720 vibrations, and adopting the twenty-seventh partial we have D, 810 vibrations; ratio to C, 8 : 9. Adopting the twenty-eighth partial we have E♭, 840 vibrations; ratio to C, 6 :7. Adopting the thirty-second partial, 960 vibrations, we have F; ratio to C, 3 : 4. Adopting the thirty-sixth partial we have G; ratio to C, 2 : 3. Adopting the thirty-eighth partial we have A♭; ratio to C, 12 : 19. Adopting the forty-fifth partial we have B♮; ratio to C, 8 : 15. The forty-eighth partial gives the octave to C below it.

THE HARMONIC MINOR SCALE.

Ordinal number of partials......	24	27	28	32	36	38	45	48
Inter. ratios......	C $\frac{18:9}{204}$	D $\frac{27:28}{63}$	E $\frac{7:8}{231}$	F $\frac{8:9}{204}$	G $\frac{18:19}{94}$	A $\flat\frac{38:45}{292}$	B $\natural\frac{15:16}{112}$	C
Pitch number	720	810	840	900	1080	1140	1350	1440
Ratios from C....	1	8 : 9	6 : 7	3 : 4	2 : 3	12 : 19	8 : 15	1 : 2
Cents from C	0	204	267	498	702	796	1088	1440

The harmonic scale provides an ascending minor scale, with a major sixth, as in the just minor scale.

THE HARMONIC MINOR SCALE, WITH MAJOR SIXTH.

24	27	28	32	36	40	45	48
C	D	E $\flat$	F	G $\frac{9:10}{182}$	A $\natural\frac{8:9}{204}$	B $\frac{15:16}{112}$	C
1	8 : 9	6 : 7	3 : 4	2 : 3	3 : 5	8 : 15	1 : 2
0	204	267	498	702	884	1088	1200

In the above harmonic minor scale the only difference from the just minor scale with a major sixth is the third.

The harmonic scale also provides a descending scale with a minor sixth and sub-minor seventh reckoned from the tonic below.

HARMONIC MINOR SCALE WITH SUB-MINOR SEVENTH.

24	27	28	32	36	38	42	48
C	D	E $\flat$	F	G $\frac{18:19}{94}$	A $\flat\frac{19:21}{173}$	B $\flat\frac{7:8}{231}$	C
1	8 : 9	6 : 7	3 : 4	2 : 3	12 : 19	4 : 7	1 : 2
—	204	294	498	702	796	969	1200

So that it will be seen that the harmonic scale provides for the just major scale, and for certain kinds of minor scales, but does not sanction in our scales the ratio of 5 : 6 used out of its proper place in exchange for 6 : 7, nor ratio 5 : 8 used out of its proper place. The history of the interval 5 : 6 shows what opposition there was to the use of it as the mediant in the minor scale, but the harmonists prevailed, and hence the addition of polyphonic to homophonic music in minor as well as major scales. Although we may thus obtain three minor scales true to nature, they are obtained by adopting in one scale the ratio 6 : 7 instead of 5 : 6 and 12 : 19 for 5 : 8 ; in another the ratio of 6 : 7 for 5 : 6 ; two series of sounds which do not form so many consonances as by adopting what is now called the just minor scale. For instance, by lowering the E♭ from ratio 5 : 6 to ratio 6 : 7, we get only 267 cents for a minor third from c, whilst we get 435 cents for the major third between E♭ and G, $i.e.$, a major third below G, too sharp by 49 cents.

It has, therefore, been wisely determined to adopt the ratio 5 : 6 and 5 : 8 for minor scales in order to allow of the notes in our musical scale forming consonances with each other.

The provision made by the harmonic scale for formulating the just major scale has, we believe, not hitherto been proved, though a similarity has been shown to exist between the two scales. This accounts, perhaps, for the statement by the writer on Harmonics in Grove's Dictionary, that, in *practically* deducing the diatonic scale (by which he evidently means the just major scale) from the harmonic scale, the intervals must be corrected by the ear. Happily, we are enabled to formulate our diatonic scales without consulting that very fallible guide the ear.

Although the writer referred to makes the statement quoted above, he undertakes to show that part of the just major scale of C, viz., from dominant to dominant, may be *scientifically* deduced from the partials of G as the base of the harmonic scale. Unfortunately, he succeeds only in showing it cannot be so deduced.

The harmonics or partials of G (starting as the writer does) on the 4th partial, as shown by the accompanying diagram, are

4th	5th	6th	7th	9th	11th	13th
G	B	D	F	A	C	E

The writer then says—taking the above notes and combining and transposing them (which means bringing them within the compass of an octave) the scale of C from dominant to dominant is formed. This is a mistake. Starting from G the 4th partial, A is acquired by lowering A the 9th partial, an octave. Its ratio to G is then

8 : 9. B is the 5th partial, ratio from A, 9 : 10.
C is acquired by lowering C the 11th partial, an
octave. Its ratio to B is then 10 : 11. D is the
6th partial. Its ratio to C is 11 : 12. E is acquired
by lowering E the 13th partial, an octave. Its
ratio to D is then 12 : 13. F is the 7th partial.
Its ratio to E is 13 : 14. G is the 8th partial. Its
ratio to F is 14 : 16 = 7 : 8.

The scale of dominant to dominant upwards,
in the key of C, according to this way of formulating
it, would be

$$G^{8:9}_{204} \mid A^{9:10}_{182} \mid B^{10:11}_{165} \mid C^{11:12}_{151} \mid D^{12:13}_{139} \mid E^{13:14}_{128} \mid F^{7:8}_{231} \mid G$$

On comparing this with the ratios and cents of the
just scale, which, from dominant to dominant, are as
follows—

$$G^{9:10}_{182} \mid A^{8:9}_{204} \mid B^{15:16}_{112} \mid C^{8:9}_{204} \mid D^{9:10}_{182} \mid E^{15:16}_{112} \mid F^{8:9}_{204} \mid G$$

it will be seen how erroneous is the statement that
by combining and transposing G B D F A C E (the
respective 4th, 5th, 6th, 7th, 9th, 11th, and 13th
partials of G) we get the just major scale of C,
ascending from dominant to dominant. Another
statement is certainly well grounded, that there is
the relationship between the two scales under consideration, of dominant to tonic; but with G as the

M

base of the harmonic scale it is wrong to allude to C (one of the partials of that scale) as illustrative of that relationship. Tonics and dominants must be related by perfect fourths and fifths, but the perfect fourth with G as base lies between D and G (ratio 3 : 4) and not between G and C (ratio 8 : 11), as is inferred, consequently G as base cannot be regarded as dominant and C as tonic.

The relationship of dominant and tonic exists only between the two scales when C is considered as tonic and G as dominant, for then only is the relationship of tonic to dominant perfect.

In our diagram F is tonic, C is dominant, which is the same as if C were tonic and G dominant.

We repeat that the provision made by the harmonic scale for the just scale is, by its partials, 24, 27, 30, 32, 36, 40, 45, and 48, and not as pointed out by the writer on Harmonics in Grove's Dictionary.

CHAPTER XVI

MEAN TONE SYSTEMS.

NO sooner were the principles on which the just scale is founded recognised to be correct, and adapted alike to the demands of science and æsthetic taste, than the impossibility was perceived of tuning instruments with only twelve fixed notes within the octave, in such a manner as to adhere to those principles in setting up eleven other scales having their initials on the notes established by the just scale and on the intermediate notes of that scale, for the sounds of the just prime scale would not furnish a correct series for a just scale commenced on any other note. Thus, if the just scale were set up on C, and then the scale of G were set up, the A of the scale of C would be too flat by 22 cents for a .major second to G. If the scale of D were set up, E of the scale of C would also be too flat by 22 cents for a major second to D; and the more distant the relations of the scales by fifths and fourths the fewer notes would the prime just scale supply which would be in tune. The question then arose whether it would not be better to sacrifice the most perfect intonation, in regard to instruments with fixed tones, to an arrangement which would allow of a certain number of scales to approximate to correct tuning, and this, even, if for such instruments the number of keys that could be played in without much annoyance to the ear should be limited.

The question being decided in favour of making a limited number approximate to correct tuning, various schemes were suggested to temper a certain number of scales. In almost all these schemes the plan was to reduce the major tone, ratio 8 : 9, cents 204, half a comma (about 11 cents), or to raise the minor tone, ratio 9 : 10, cents 182, half a comma, which amounted to the same thing. This was taking the arithmetical mean of two intervals, and hence the schemes for sacrificing true intonation were called *mean tone systems.* In some systems the third note was flattened, in others the fifth was slightly or considerably flattened; and there was one system in which the measure was that of a divided tritonis (or interval of two minor and one major tones); but the system which was generally adopted resulted in making the second a mean tone second (cents 193), in making the third a perfect major third (cents 386), in raising the fourth to 503 cents, in flattening the fifth to 696 cents, in raising the sixth to 889 cents, in flattening the seventh to 1082 cents, and in dividing the *mean tone sound* so as to give 76 and 117 cents, with 117 between the mediant

and subdominant and 117 cents between the leading
note and octave to the tonic.

THE INITIAL SOUNDS OF TWELVE SCALES, WITH MEAN TONE TUNING.

$$C_{76} \mid C^{\sharp}_{117} \mid D_{117} \mid E^{\flat}_{76} \mid E^{\natural}_{117} \mid F_{76} \mid F^{\sharp}_{117} \mid G_{76} \mid G^{\sharp}_{117} \mid A_{117} \mid B^{\flat}_{76} \mid B^{\natural}_{117} \mid C$$
$$76 \quad 193 \quad 310 \quad 386 \quad 503 \quad 579 \quad 696 \quad 772 \quad 889 \quad 1006 \quad 1082 \quad 1199$$

MEAN TONE SCALES, WITH THE WIDTH OF THE INTERVAL DEFINED BY CENTS.

A Mean Tone Scale of C.

$$C_{193} \mid D_{193} \mid E_{117} \mid F_{193} \mid G_{193} \mid A_{193} \mid B_{117} \mid C$$
$$193 \quad 386 \quad 503 \quad 696 \quad 889 \quad 1082 \quad 1199^{*}$$

A Mean Tone Scale of D.

$$D_{193} \mid E_{193} \mid F^{\sharp}_{117} \mid G_{193} \mid A_{193} \mid B_{193} \mid C^{\sharp}_{117} \mid D$$
$$193 \quad 386 \quad 503 \quad 696 \quad 889 \quad 1082 \quad 1199$$

A Mean Tone Scale of F.

$$F_{193} \mid G_{193} \mid A_{117} \mid B^{\flat}_{193} \mid C_{193} \mid D_{193} \mid E_{117} \mid F$$
$$193 \quad 386 \quad 503 \quad 696 \quad 889 \quad 1082 \quad 1199$$

A Mean Tone Scale of G.

$$G_{193} \mid A_{193} \mid B_{117} \mid C_{193} \mid D_{193} \mid E_{193} \mid F^{\sharp}_{117} \mid G$$
$$193 \quad 386 \quad 503 \quad 696 \quad 889 \quad 1082 \quad 1199$$

A Mean Tone Scale of A.

$$A_{193} \mid B_{193} \mid C^{\sharp}_{117} \mid D_{193} \mid E_{193} \mid F^{\sharp}_{193} \mid G^{\sharp}_{117} \mid A$$
$$193 \quad 386 \quad 503 \quad 696 \quad 889 \quad 1082 \quad 1199$$

A Mean Tone Scale of B♭.

$$B^{\flat}_{193} \mid C_{193} \mid D_{117} \mid E^{\flat}_{193} \mid F_{194} \mid G_{193} \mid A_{117} \mid B^{\flat}$$
$$193 \quad 386 \quad 503 \quad 696 \quad 889 \quad 1082 \quad 1199$$

* The loss of 1 cent, 1199 for 1200, is immaterial.

We give below a diagram of the just major scale, in order to facilitate comparison with the preceding mean tone system.

THE JUST SCALE.

C_{204} | D_{182} | E_{112} | F_{204} | G_{182} | A_{204} | B_{112} | C
204 | 386 | 498 | 702 | 884 | 1088 | 1200

The octave divided into semitones, as shown in the previous diagram, yielded only six scales with like intervals, and consequently musicians avoided (because still more out of tune) all the other scales and keys as much as possible, or only adopted them on the harpsichord or clavecin, rapidity of execution upon which prevented the ear from being distressed with the false relations of the tones.

CHAPTER XVII

THE EQUAL TEMPERED SCALE.

THIS scale owed its construction to the desire of musicians to be afforded the means of playing upon instruments with fixed tones in all and each of the keys within the octave; also to modulate from one key to another, without producing that distressing phenomenon called "beats" which result from the simultaneous sounding of tones not justly related, or not tuned in exact consonance with each other.

Under the mean tone system, performances on instruments with only twelve tones were limited to six keys. Musicians who claimed to be considered as "progressionists" rebelled against this restriction, and determined to favour a system which should overcome the difficulty in question.

To obtain the long-sought for desideratum, and extend the opportunity for playing in every key to every instrument, it was determined to divide the octave into twelve equal parts, or, in other words, to consider each part as one-twelfth of the octave. This idea corresponds with Mr. Ellis's plan of reckoning by 1200 cents, and subdividing the numbers appropriated to each interval until the full tune is reached by the last note of the octave. The difference in Mr. Ellis's plan and that developed in the equal-tempered system, is, of course, apparent. The subdivisions of the just scale vary with the width of the intervals. Those of the equal-tempered system allow 100 cents to each semitone and 200 cents for each whole tone, the entire twelve parts being represented by 100 cents to each part. The following diagram represents a *Chromatic Scale* formed upon this plan :—

C100 C♯D♭100 D100 D♯E♭100 E100 F100 F♯G♭100

CHROMATIC SCALE CONTINUED—

G100 G♯A♭100 A100 A♯B♭100 B100 C.

DIATONIC SCALE.

	C200	D200	E100	F200	G200	A200	B100	C
Cents	0	200	400	500	700	900	1100	1200

Thus, whatever note the scale is founded on, as the initial tone, every scale is equally out of tune.

THE RATIOS OF THE EQUAL-TEMPERED SCALE.

These, reckoned from a prime, are as follows :—

400:449, 50:63, 227:303, 259:433, 22:37, 89:168

If the long-cherished theory of both ancients and moderns be true—namely, that the beautiful in sound is dependent upon a series of intervals defined by simple ratios or whole numbers—then we can scarcely find, either in ancient or modern musical history, a more striking example of departure from this rule than that presented by the equal-tempered system—one, in fact, which exhibits no single interval which is defined by simple ratios.

The compensative part of this system is to be found in the opportunity which it affords for instruments with fixed tones to play in every key instead of the limited few, whilst the actual departures from the tonal perfectness, attained by the subdivisions of the just scales, although distressing to highly-sensitive ears, are not so great as to impinge painfully upon those which cannot appreciate perfectly true intonation.

SECTION II.

CHAPTER I.

THEORIES CONCERNING THE CAUSE AND EFFECT OF THE BEAUTIFUL IN SOUND.

THE musical scientist of modern times has not been satisfied to pause upon the discovery that the coalescence of vibrations, indicated by figures denoting simple ratios, such as $1 : 2$, $2 : 3$, $3 : 4$, &c., are the representatives of the beautiful in sound.

With each fresh advance into the realm of musical science inevitably arises a new series of questions involving psychological as well as mathematical problems, and the demand to know why intervals measurable by simple ratios should be more acceptable to the cultured ear than those determined by complex ratios will force itself upon the attention of the reflective student. And this is not all— scarcely, indeed, the beginning of the urgent questioning that must arise when a science, which, *à priori*, seems to appeal simply to the taste and fancy of the listener, begins to range itself under the domain of immutable law, and the most eminent leaders of musical art enunciate the startling propositions, "thus must it be," and, "that cannot be allowed," in order to produce the beautiful in sound. As early as the middle of the eighteenth century, the theory propounded on this subject by the celebrated mathematician Euler, gained much favour with those who were desirous of finding a philosophical source for the delightful sensations caused by certain forms of music, as they impinged upon the auditory nerves.

Euler contended that the mind is pleased with everything in which we can find a certain amount of perfection—and again, that perfection in everything is determined by the co-operation of all parts towards the attainment of its end. Thus, the learned philosopher argues, in his published treatises, that "perfection arises from order—order being an arrangement of all parts in the special place they should occupy," &c. "Whilst," he adds, "the mind delights in order, the relations of the several parts must be sufficiently simple and apparent to be readily perceived. Hence the superiority of the simple ratios of music over the complex, and hence a combination of tones will please us, when we can discover the law of their arrangement."

In commenting on this theory, Helmholtz justly remarks, "The principal difficulty in Euler's theory is, that it says nothing of the mode in which the mind is enabled to perceive the numerical ratio of two combined tones. "A man left to himself is scarcely aware that a tone depends upon vibrations. Moreover, immediate and conscious perception by the senses has no means of discovering that the number of vibrations performed in the same time are greater for high than for low tones, and that determinate intervals have determinate ratios. . . . A man that has never made physical experiments, or who has never had the opportunity of knowing anything about vibrational numbers or their ratios and vast numbers who still remain ignorant on these points all their lives, may yet experience the keenest delight in music."

It must be added that Helmholtz does not rest satisfied with emphasizing the weak points of Euler's theory, but he offers certain deductions, based upon scientific principles, which appeal with far more force to the analytical mind than the propositions of Euler. Whilst the latter affirms that the gratification produced on the mind by music proceeds from an intelligent perception of the vibrational order in which music is arranged, Helmholtz, on the contrary, refers the delight experienced by the mind, to the *physical effect of that same vibrational arrangement of tone upon the ear.* He says, "The human mind perceives only the *physical* effect in either the continuous or intermittent sensations of the auditory nerves."

In attempting to demonstrate this proposition, Helmholtz concludes an exhaustive series of arguments, by affirming that " a continuous, and therefore pleasurable sensation is produced on the nerves of audition, when there are no disturbing *beats* to interfere with the continuity of the sounds, but that an intermittent, and therefore disturbing sensation, arises from the action of *beats*."

To apprehend either the truth or value of this theory, it is necessary to become acquainted with a set of tonal phenomena apprehended in general detail by modern musicians, but so elaborately treated of by Helmholtz that we shall adopt his nomenclature in describing them, and his method of ranging them under the designation of " Partials, " Combinational Tones," and " Beats."

To render each department of this subject as intelligible as possible, it will be necessary to sum up what is now generally known concerning vibratory action in sound.

The study of acoustical laws shows, first, that the oscillations of the particles of any resonant body increase or diminish in proportion to the force with which the sound is generated; secondly, when we strike a string, its vibrations or oscillations are sufficiently palpable to be perceived by the eye, and as the first impulse gives the loudest sound, so as the tone diminishes the *visible* vibrations become less and less perceptible; thirdly, the same results are obtained when we increase our distance from the sounding body in the open air, for it is only the amplitude of the oscillations of the air particles which diminishes, as their distance from the sounding body increases. Hence, loudness depends on this amplitude, whilst pitch depends solely on the length of time in which each single vibration is executed, or on the *vibrational number* of the tone.

In these calculations we take a second of time as the unit, and consequently imply by the term "*pitch*" or "*vibrational number*," the number of vibrations given off by the particles of a sounding body in one second of time. Thus, it is absolutely proved, that musical tones are more acute, or higher, the greater the number of vibrations; hence, the quicker must be the period of those vibrations performed in a second of time.

It would be superfluous in this place to attempt enumerating the various mechanical contrivances that have been invented for registering and calculating vibrations. "The Siren" (after Seebeck, Dove, and others), Sir John Herschel's admirable instruments, besides many of still more recent date, have enabled modern musicians to calculate, even to the thirty-thousandth part of a second, the number of vibrations produced by each musical tone. Still, it must be obvious that the nerves of the human ear are only susceptible of receiving impressions within a scale far too limited to bear any proportion to an infinitesimal number of vibrations ; for although we are justified—by the known laws of motion—in believing in the endless influence of vibratory action when once set up, the powers of audition in the human organism, even in their acutest extent, are too well known, and their limitations too clearly defined, to admit of any possibility of their apprehending the integral units of a vibrational tone.

The lowest tones at present adopted for orchestral instruments are the $E_{\prime}$ of the double bass, the result of forty-one and a quarter vibrations in a second ; and $C_{\prime}$ of the double bassoon, the result of thirty-three vibrations in a second. The grand pianos of recent date produce $A_{\prime\prime}$ with twenty-seven and a half vibrations. Still, it has been found that the pitch of all musical tones below $E_{\prime\prime}$ with forty-one and a quarter vibrations, are so difficult to distinguish that they can only be used with effect for strengthening their octave note above.

The same limitations of the auditory nerves are applicable to high notes which are the results of many thousand vibrations in a second.

Practical experiment has proved that the

musical tones which can be used with advantage, and have a distinguishable pitch, are those which range between 33 and 4,752 vibrations in a second. Still, tones are audible, and their pitch appreciable, when they are caused by 42,240 vibrations in a second. We say *appreciable* rather than *distinguishable*, a much smaller range of tone being required to produce notes that will impinge upon the ear with a direct intimation of a clear ringing musical tone.

As an example of vibrational numbers, corresponding to different degrees of pitch, we give the following scale of tones in a modern pianoforte, ranging from $C_{\prime}$, with thirty-three vibrations, to the C'''', the 8th octave above—

C	2nd oct.	3rd oct	4th oct.	5th oct.	6th oct.	7th oct.	8th oct.
33	66	132	264	528	1056	2112	4224

though the highest note of the piccolo is the result of 4,752 vibrations, but M. Appunn has succeeded in producing a note four octaves and a minor tone above the piccolo's highest note, his experiments being made by a number of tuning-forks excited by a violin bow.

It may be understood from even this brief notice of vibratory action in musical sounds that the more we advance into the realm of acoustical laws the more certainly may we realise that the underlying causes of that gratification which the ear derives from music is a result of the perfectly equal times of the vibrations that each tone produces.

It should be observed that two of the subjects treated of in the following pages have been known to modern musicians under the familiar appellation of "Harmonics." Students of the recondite principles discoverable in vibratory tonal action will recognise the propriety of classifying the phenomena in question under Helmholtz's nomenclature of " Partials," " Combinational Tones," " Coincidences," and " Beats."

Chapter II

HARMONIC PARTIALS.

TO apprehend clearly what is implied by the term "Partials" it must be premised that any elastic body, such as a stretched string, a resonant piece of metal, &c., when struck by an adequate force, not only vibrates as a whole, but that a similar vibratory action ensues amongst and within the aliquot parts of the body in question. Thus in a string of a given length there are vibrations arising from the half, the third, and other aliquot parts of the string, whilst the full tone is composed of the sum of all these vibrations, or the combined tones of many aliquot parts.

"It is well known," remarks Helmholtz, "that the union of several simple tones into one compound tone is artificially imitated in the organ by special mechanical contrivances," and musicians must regard the tones of all musical instruments as being a similar combination to the compound stops of an organ.

The aliquot parts of a compound tone, generally called "Harmonics," have been more correctly designated by Helmholtz as Upper Partials.

In examining the effects produced on the ear by these upper partials, we encounter some strange phenomena, known to and recognised by physicists, but treated more as a curious phase of acoustics than as a significant indication of tonal laws.

A priori it is known that if the ear is properly directed to the vibrations which occur when a full rich musical tone is produced, the ensuing vibratory action does not give off one tone alone—*i.e.*, that of the prime—but a whole series of ascending tones are heard, a series which never varies, whatever may be the generating note, and which follows the order given in the succeeding diagram.

1	2	3	4	5	6	7	8	9	10	11	12	13	14	15	16
C	c	g	c′	e′	g′	bb′	c″	d″	e″	f″	g″	a″	bb″	b″	c‴
66	132	198	264	330	396	462	528	594	660	726	792	858	924	990	1056

The upper line of figures shows the series of partials up to the number 16. Any pair of the same figures, for example 1 and 2, 2 and 3, indicate the ratios of the underlying sounds, and the pitch numbers or number of vibrations causing those sounds when the prime is due to 66 vibrations in a second of time.

On referring to the above scale it will be seen that the prime note C must be regarded, in itself, as a *partial*, to which the second partial, or first *upper* partial, is its octave c, making double the number of vibrations to the prime tone and first partial. The second *upper* partial, or third partial,. is the twelfth, with three times as many vibrations. as the prime. The third *upper*, or fourth partial is the second octave to the prime, with four times as many vibrations. The fourth *upper*, or fifth partial, is the major tenth, with five times as many vibrations as the prime. The fifth *upper*, or sixth partial, has six times as many vibrations as the prime. The sixth *upper*, or seventh partial, makes seven times. as many vibrations as the prime, whilst all the other upper partials correspond in the same regular scale of increase in their number of vibrations. The ratios, in regard to the pitch numbers, however,. are not all of a simple kind, consequently, if consonance (or the agreement of sounds) depends on simplicity of ratio, even a compound sound, unless it be composed of six partials only, must include the elements of dissonance. For example—with C as a prime tone we have for the eight upper partials— *c*, *g*, *c'*, *e'*, *g'*, *bb'*, *c"*, *d"*, all combining to form the compound of the prime C, but comprising dissonances from the ratios of 1 : 7, 1 : 9, besides the dissonances which arise between some of the first six and some of the higher upper partials.

To obtain a complete demonstration of these upper partials, and arrive at the certainty that tones are compounded of the vibrations noted above, a full rich generating prime tone is necessary. Helmholtz suggests the following means as one amongst many other available modes of obtaining the desired result. He says—

"Touch the node of a pianoforte string or monochord with a camel-hair pencil; strike the string, and immediately remove the pencil from the string. If the pencil has been pressed tightly on the string we either continue to hear the required partial as an harmonic, or else in addition hear the prime tone gently sounding with it. On repeating the excitement of the string, and continuing to press more and more lightly with the pencil, and at last removing it entirely, the prime tone of the string will be heard more and more distinctly with the harmonic till we have finally the full natural tone of the string. By this means we obtain a series of gradual transitional stages between the isolated partial and the compound tone, in which the first is readily retained by the ear. By applying this last process I have generally succeeded in making perfectly trained ears recognise the existence of upper partial tones."

Upper partial tones, even if they are not appreciable to the ordinary or indifferent listener, are nevertheless always present in the sensation which a rich musical tone produces on the ear.

Their combined effect has not inaptly been likened to the taste of a well-prepared mass of culinary condiments wherein the several ingredients may not be distinguishable in detail, whilst yet the *toute ensemble* forms the triumph of the *chef's* art.

Still more pertinent to the subject is the effect produced on the retina of the eye by a beam of what we are accustomed to speak of as pure white light. Subject this beam to the disintegrating action of a prism, and we find that the one beam is but a combination of seven variously coloured strands, and the assumed purity or simplicity of the white ray is a subtle compound of all the lustrous shadings displayed in the rainbow.

And these similes will not appear to be out of place if we consider the existence of "harmonics," or, as we prefer to call them, after Helmholtz, "upper partials," which are as much a fixed fact in tonal principles as the seven prismatic colours in the constitution of a beam of white light. To enforce the necessity of treating this subject with the most profound and searching analysis, we shall close our notice of upper partials (as preliminary to treating of the phenomena connected with them) by again citing the opinion of Helmholtz, whose researches on the more abstract principles of tonal laws entitle him to be regarded as a standard authority. This writer says*—

---

* Ellis's Translation.

" We have thus been led to an appreciation of upper partial tones which differs considerably from that previously entertained by other musicians and even physicists, and must therefore be prepared to meet the opposition which will be raised. The upper partial tones were indeed known, but almost only in such compound tones as those of strings where there was a favourable opportunity for observing them; but they appear in previous physical and musical works, as an isolated accidental phenomenon of small intensity; a kind of curiosity which was occasionally adduced, in order to give some support to the opinion that nature had prefigured the construction of harmony, but which, on the whole, remained almost entirely disregarded. . . . In opposition to this we assert—that upper partial tones are, with a few exceptions already named, a general constituent of all musical tones, and that a certain stock of upper partials is an essential condition for a good musical quality of tone. Finally, these upper partials have been erroneously considered as weak, because they are difficult to observe, while, in point of fact, for some of the best qualities of tone, the loudness of the first upper partials is not far inferior to that of the prime tone itself."

We shall now refer to the second phenomenon alluded to by Helmholtz, namely, a vibratory action resulting from the combination of two tones of different pitch, the effect of which must of necessity differ materially from the upper partials produced by a single prime tone.

COMBINATIONAL TONES.

Helmholtz prefaces his description of "Combinational Tones" with the following remarks :—

"Oscillating motions of air produced by several sources of sound acting simultaneously are always the exact sum of the individual motions of each sound separately . . . and this law is a most important one, for it reduces the consideration of compound tones to simple ones. . . . But the law holds strictly only in the case where the vibrations in all parts of the mass of air and of the sonorous elastic bodies are of infinitesimally small dimensions. In all practical application of this law to sonorous bodies the vibrations are always very small, and near enough to being infinitesimally small, for the law to hold with great exactness even for the real sonorous vibrations of musical tones, and by far the greatest part of this phenomena can be deduced from that law in conformity with observation."

Still there are certain phenomena which result from the fact that this law does not hold good with perfect exactness for the vibrations of elastic bodies which though almost always very small are far from being infinitesimally small.

One of the phenomena with which we are now interested is the occurrence of "*combinational tones,*" which were first discovered in 1745 by Lorge, a German organist, and were afterwards generally known—although a wrong pitch was assigned them—to the Italian violinist Tartini (1754), from whom they were termed Tartini's tones. Helmholtz adds :—

" These tones are heard whenever two musical notes of different pitch are sounded together loudly and continuously. The pitch of a combinational tone is generally different from either of the gene-- rating tones or their upper partials. . . . In conducting experiments, the combinational tone may be readily distinguished from the upper partials by not being heard when only one generating note is sounded, and by appearing simultaneously when a second note is added. . . . Combinational tones are of two kinds. The first class I have named *differential tones*—the second (discovered by myself) I call *summational tones*. The pitch number of the first is the difference of the pitch numbers of the generating tones. The second have their pitch numbers equal to the sum of the pitch numbers of the generating tones."

Before entering upon the analysis of those *differential* and *summational* tones, into which Helmholtz divides *combinational* tones we give the lettering he adopts in alluding to those sounds, and reproduce the harmonic series of sounds to facilitate reference to it.

SYSTEM OF DESCRIBING SOUNDS BY LETTERS, AS USED BY HELMHOLTZ.

The seven notes of the scale of C on the second line below of the bass clef are represented by capital letters thus—

C D E F G A B

The seven notes of the scale of C on the first line below of the treble clef, are represented by small letters thus—

$$c \quad d \quad e \quad f \quad g \quad a \quad b$$

The seven notes of the scale of C on the third space of the treble clef are represented by small letters with an accent over them thus—

$$c' \quad d' \quad e' \quad f' \quad g' \quad a' \quad b'$$

The seven letters of the scale of C on the second line above of the treble clef are represented by small letters with two accents over them thus—

$$c'' \quad d'' \quad e'' \quad f'' \quad g'' \quad a'' \quad b''$$

In order to represent higher octaves, for every additional octave add another accent to the small letters thus—c''', d''', &c.

In order to represent lower octaves than C, add one, two, &c., accents below thus—$C_{,}$ $C_{,,}$.

SCALE OF PARTIALS, OR HARMONIC SERIES OF SOUNDS.

Ordinal number of partials	1	2	3	4	5	6	7	8	9	10	11	12	13	14	15	16
Letters for sounds	C	c	g	c'	e'	g'	b♭'?	c''	d''	e''	f''	g''	a''	b♭''	b♮'	c'''
Pitch numbers	66	132	198	264	330	396	462	528	594	660	726	792	858	924	990	1056

The figures in the first line indicate the ratios of the pitch numbers, and show how many times the corresponding pitch number is greater than that of the prime C. Those in the lower line give the pitch numbers of all the notes when C gives sixty-six vibrations.

CHAPTER III

ON DIFFERENTIAL TONES.

WHAT is termed a "differential" tone may be heard—

1. When one prime tone is combined with another prime tone of different pitch, *i.e.*, E or C with G.

2. A differential tone may be heard, generated by a primary and an upper partial.

3. A differential tone may be heard generated by a differential tone and either of its generators.

1. *When one prime tone is combined with another prime tone of different pitch.*

As the pitch number of a differential tone is the difference of the pitch number of the generating tones, when *c* (vibrations 132) as a prime is combined with its fifth (*g*), the pitch of the differential tone is found by taking the pitch numbers of *c* and *g* severally, or the ratio of the vibrational numbers in the harmonic scale, which are 2 and 3. The first differential tone is therefore 1, or C (vibrations 66) the octave below the small *c*. This can also be shown thus :—

First generating tone *c*, 132 vibrations.
Second generating tone *g*, 198   ,,
Difference 1
Differential tone C   66   ,,

Again, combining c (132 vibrations) as a prime with its octave above c' (264 vibrations) the ratio of pitch number, 2 to 4, and the difference, 2, the pitch number of the differential tone, would be 2, *i.e.*, c (132 vibrations), or the same as the pitch of the lowest generator; also shown thus :—

First generating tone c, 132 vibrations.
Second generating tone c' 264   ,,
Difference 2
Differential tone c 132   ,,

2. *When the Differential Tone is generated by a Prime and Upper Partial.*

" Differential tones " may be produced not only by two prime generators but also by primary and upper partial tones.   Helmholtz says :—

" On carefully analysing the combinational tones of two compound musical tones, we find that both the primary and upper partial tones may give rise to differential and summational tones."

He adds that most commonly the differential are stronger than the summational tones, whilst the stronger generating tones also produce the stronger combinational tones, and since in musical compound tones the prime generally predominates over the partial tones, the differential tones of the primes are heard more loudly than the rest, and were consequently first discovered. These tones are more readily recognised when the two generating tones are at closer intervals than an octave apart, in which case the differential tone is deeper than in either of the generating tones.

In order to show how differential tones are generated from primary and upper partials, Helmholtz says—

"Take the major third of c', viz., e', ratio of pitch number 4 to 5. The first difference is 1, *i.e.*, C. The first harmonic upper partial of c' is its octave c'', relative pitch number 8, ratio 5 : 8, difference 3, *i.e.*, g. The first upper partial of e' is e'', pitch number 10, ratio for this and c' 4 : 10, difference 6—that is g''. Then again, $c''e''$, ratio 8 : 10, difference 2, *i.e.*, c. Hence, taking only the first upper partials, we have the series of combinational tones, 1, 3, 6, 2, or C $g\,g'\,c$."

Of these Helmholtz says the tone 3 or g is often easily perceived.

3. *When a Differential Tone is generated by a Differential Tone and one of its Generators.*

Helmholtz says—

" A multiple combination as above may be considered as arising thus : The first differential tone or combinational tone of the first order by combination with the generating tones themselves produce other differential tones, or combinations of the second order, and these again produce new ones with the generators and differentials of the first order, and so on," &c., &c.

To show how the different orders which he describes arise, Helmholtz says—

"Take two simple tones, c' and e', ratios 4 : 5, difference 1.   Differential tone of the first order C. This (that is, C) with the generators gives the ratios 1 : 4 and 1 : 5, differences 3 and 4, differential tones of the second order g and c' once more.   The new tone gives with the generators the ratios 3 : 4 and 3 : 5, differences 1 and 2, giving the differential tones of the third order C and c, and the same tone 3 gives with the differential of the first order 1 the ratio 1 : 3, difference 2.   Hence, as a differential of the fourth order, c once more, and so on."*

Helmholtz gives illustrations of the notes of multiple combinations when primes are combined of the ratios of 1 : 2, 2 : 3, 3 : 4, 4 : 5, &c.   Ellis remarks that these examples are best calculated by giving to the notes the numbers representing the partials of the harmonic scale.

---

* No new note arises when the differential note is C and the generators are c' c', for the differential note of the second order is again c'.

Chapter IV

SUMMATIONAL TONES.

SUMMATIONAL tones, according to Helmholtz, are "acute resultant sounds generated by a prime combined with another note of a different pitch." The same authority says, "Summational tones" may arise from the upper partials, but he claims special attention only for those tones arising from combined primes.

Helmholtz affirms that the pitch number of summational tones is the pitch number of the sums of the generating tones. Thus, if we inquire what is the pitch of the summational tones when c (132 vibrations) is combined with its fifth (g), we find the ratio of the pitch numbers of c and g to be 2 to 3. Thus $2 + 3 = 5$. The pitch number of the summational tone is therefore 5, and 5 is e', so that c combined with g will give rise to e' (vibrations 330).

Again, if we inquire what is the summational tone of c (132 vibrations), combined with its octave, the ratio of the pitch numbers is 2 to 4, their sum 6. The pitch number ratio is therefore 6, which is g' (vibrations 396). If we inquire what is the summational tone with two sounds, say g (198 vibrations) and c' (264 vibrations) combined, we find the pitch numbers are 3 and 4, and their sum 7, the pitch number 7 is bb', consequently g and c' combined produce bb' (462 vibrations).

In the same way it can be shown that the summational tone, when g and e' are combined, is c'' (528 vibrations), and when the major third c' and e' are combined it is d''. When the minor third e' and g' are combined it is f''. When the minor sixth, e' and c'' are combined it is a''.

Diagram of Summational Tones said to be produced by Two Primes.

Summational Tones	g'	e'	bb'	c''	d''	f''	a''
Generator	c'	g	c'	e'	e'	g'	c''
Generator	c	c	g	g	c'	e'	e'

(Read the columns upwards.)

Some of these summational tones form inharmonic intervals with the generators, and were they not but faintly heard on most instruments they would produce the effect of dissonance.

The influence of Combinational Tones on the Harmoniousness of Two Sounds.

In considering the effect which combinational sounds are shown to produce in relation to their generators, it will be observed that the *continuity of sound*, so desirable in pure harmony, must of

o

necessity be interfered with by their action. Differential tones may be said to reinforce the harmonious
accordance of their generators when these last are
properly related—adding, on the contrary, vibrational dissonance to generating tones not har-
moniously related. In respect to summational
tones, our illustrations clearly show that their
tendency is as frequently unfavourable as
otherwise.

Whether multiple combinations give rise to
differential sounds, or whether summational tones
are distinct from partial tones, are questions which
at present are far from being satisfactorily decided.

Mr. Ellis considers that the existence of differential tones of *higher orders* are not completely
established, and in his notes to the appendix of
Helmholtz's *Sensations of Sound*, he alleges that the
whole subject of combinational tones, beats, &c.,
require much farther investigation before well
defined theories can be presented for acceptance.

The question whether combinational tones are
really objective—*i.e.* arise from waves of air outside
the organism of the ear—or subjective, and formed
through the organism of the ear, has only been
settled in relation to differential tones, and the
verdict of scientists is that these tones at least may
be regarded as objective.

In the present stage of acoustical science, especially in the absence of such accurate mechanical means of analysing the compound characteristics of musical tones as would enable us to bring positive demonstration to the aid of plausible theory, we are in no position to endorse or refute the suggestions advanced by Helmholtz, especially as his theories are deduced from profound research and long-continued investigation of the tonal subtleties of which he treats.   There are, however, certain broad analogical propositions in the natural sciences which must hold their due relation to tonal as to all other laws.   Amongst these it must be remembered that combination amongst various elements invariably tends to change their original character.

Throughout the whole range of chemical affinities, for example, it is an admitted fact that two primaries in combination produce a proximate different from either of the originals, whilst the union of two different proximates produces an ultimate of a character no less differential.

What varied results, therefore, may we not look for in proportional combinations of atmospheric vibrations ?

The simple culinary feat of converting a table-spoonful of the albuminous matter contained in an eggshell into a gallon of fair white froth, by the

mere process of beating it, is but a familiar exemplification of the magical transformations to be effected
by motion.

Why may not the same laws of motion in solids,
and those which govern chemical affinities, hold good
in the combinations of tone effected by different
currents of atmosphere stirred by different numbers
of vibrations ?

When dealing with the extreme attenuations of
tone which upper partials produce, especially when
the subject for experiment is to be a greater or less
number of combined tones, it might be difficult, and
even impractical, to follow Professor Helmholtz to
the extent of assigning titles, and recording definite
numbers, for the different combinations of tone
alleged to exist in his theories.

That vibrational action does exist, and is set in
motion by musical sounds, until that action proceeds
and perhaps duplicates itself *ad infinitum*, is an
admissible proposition ; but whether the acutest of
auditory sensations can realise this action, is a point
we are not prepared to prove. Meantime, the upper
partials of a *single* generating tone, by giving off a
succession, in their variety of pitch, prove that they
are each and all component parts of that generating
tone. Still no conclusive demonstration exists to
show that the harmonics of a number of combined

tones preserve their separate individuality. Far more reasonable is it to suppose, that a combination of upper partial tones may follow the laws of affinity in the same manner as the particles of other elastic bodies. Thus, then, they may not only combine, but in their combinations produce proximate and ultimate tones of a wholly different character from the harmonic upper partials of a single generating tone.

Leaving for the present such propositions as must be considered too speculative to pronounce upon with scientific precision, we turn to another branch of our subject, and one more susceptible of definition and practical demonstration, namely, those phenomena of tonal laws called "beats."

CHAPTER V.

ON BEATS.*

TO define the nature of "beats" in the simplest possible way, we may describe them as interferences which affect the sense of audition in the form of a jar, slight impulse, or throb, in the flow of tone.

Beats are generated amongst vibrations, and occur when prime tones and upper partials are not duly related to each other, *i.e.* cannot be described by simple ratios.

The phenomena of beats arise under one or more of the following conditions :—

1. When the vibrations of a prime are disturbed by those of not simply related upper partials.

2. When primes are not simply related.

3. When the partials of two simply related primes are not simply related.

4. When the vibrations of combinational tones disturb other sets of vibrations.

1. *On Beats from the Vibrations of a Single Prime.*

As any full toned prime is known to be compound, or made up of the aliquot parts

---

* Beats only occur when one set of vibrations is not duly related to another.

represented in the harmonic scale, so beats may arise amongst the upper partials not related to the prime by simple ratios. Thus a prime may produce beats with its seventh and ninth partials.

2. *On Beats from Two Primes.*

When two primes not related to each other by simple ratios are combined, beats are produced— in fact, they may, and do, occur between all tones not duly related to each other, such, for example as are given off by instruments not properly tuned, or between any two sounds the interval of which is not consonant.

When primes are not properly related, as, for example, when a prime C is combined with its minor or major seventh; or its minor or major ninth, besides beats arising from the want of proper relationship between the primes, beats occur between the B♭ and the second partial of the prime C, which is its octave. When C *prime* is combined with B♮, beats occur not only between these two unrelated notes but also between the second partial of the first prime and the first partial of the second prime. Beats occur when C is combined with either D♭ or D♮, because, not only as in the case of C and B♭ or B♮, the intervals are too close to be described by simple ratios, but also because the second partial of C disturbs the vibrations of both D♭ and D♮.

3. *On Beats by the Partials of Two Primes properly related.*

Beats occur between the partials of two primes when their vibrations do not coalesce, as described by the ratio 1 : 1. Two properly related primes do not generate beats,* but beats may occur amongst their partials. For example, although the perfect fifth C and G (ratio 2 : 3) do not *as primes* generate beats, the third partial of G being D, is only distant from the fourth and fifth partial of C by the interval of a major second, hence the occurrence of beats.

N.B.—When the primes are not properly related to each other, all the partials of one prime create beats with the partials of the other prime.

4. *On Beats by Combinational Tones.*

When two or more prime tones are sounded at the same time beats may arise from the combinational tones as well as from the upper partials.

The loudest combinational tone is that corresponding to the difference of the pitch numbers of two primes, viz., a *differential tone of the first order.* "This is the combinational tone," says Helmholtz, "which is chiefly influential in producing beats." Helmholtz adds that "The *differential tones of the*

---

* Except, of course, those which every single compound note may generate.

first order alone, and independently of the combinational tones of higher orders, are liable to cause beats."

(*a*) Beats by combinational tones, differential and summational, arise not only when two compound tones are sounded together, but when *simple* tones are sounded together.

(*b*) Beats are also generated by the combinational tones of combinational tones.

1. *On the Beats of Differential Tones of Compound Sounds.*

In the same way as prime tones in most cases develope combinational tones, it has been shown that any pair of upper partials of the two compounds, may develope combinational tones, and it is when one or more of these combinational tones do not perfectly accord with the others, and with all the primes, and their upper partials, that beats ensue. Helmholtz illustrates this position by pointing out that a fifth tuned somewhat incorrectly, and having the pitch numbers 200 and 301, instead of (when correctly tuned) 200 and 300, must have vibrating partials which generate combinational tones, the vibrations of which differ by 2, 4, 6, and hence produce 2, 4, 6 beats, in the same time that the two first partials produce two beats. Helmholtz adds, that beats do not arise from differential tones of the first order, in the case of the generators being compounds, when there are no beats from upper partials, for he emphasises the remark "that the first differential tones of compounds only generate beats when the upper partials of the same compounds generate them."

2. On Beats by Summational Tones which disturb the Harmoniousness of Compound Tones.

Summational tones produce beats when they are not related to their generators by simple ratios. For instance, the summational tone of c' and e' being the sum of 4 and 5 is d'', and the interval between c' and d'' being a major ninth, and the interval between e' and d'', being a minor seventh, beats arise.

3. On Beats by Combinational Tones when the Generators are Simple Sounds, i.e., that is, do not develope Partials.

The combinational tones of two *simple* sounds may give rise to beats. Helmholtz argues for this position by saying—

"If combinational tones were not taken into account, two simple tones could not produce beats unless they were very nearly of the same pitch. . . . For any large intervals between two simple tones there would be absolutely no beats at all if there were no upper partials or combinational tones. . . . In fact, there would be no distinction at all between wide consonant

intervals and absolutely dissonant intervals. . . .
Now, such wide intervals between simple tones are
known to produce beats, so that even for such tones
there is a difference between consonances and
dissonances, although it is much more imperfect
than for compound tones. And these facts depend—
as Scheibler has shown—on the higher orders of
combinational tones."

Helmholtz then proceeds to argue that it is
to combinational sounds that those beats are due
which arise when two simple tones are sounded
together which do not make exact consonance.
In speaking of an imperfectly tuned third he
says :—

"A mistuned third is as pleasant, produced by
two stopped pipes,* as a just minor or major
third. This does not mean that a well-practised
ear would not find such an interval strange and
unusual, and hence would call it false ; but the
immediate impression on the ear—the simple
impression of harmoniousness considered independently of any musical habit, is in no respects worse
than for one of the perfect intervals."

Finally, Helmholtz says :—

"Collecting the results of our investigations upon
beats, we find that when two or more simple tones
are sounded at the same time they cannot go on
sounding without mutual disturbance, unless they
form with each other certain perfectly definite

---

* Stopped pipes do not develope partials.

intervals.   Such   an   undisturbed   flow   of   simultaneous tones is called a *consonance*.   When these
intervals do not exist, *beats arise*—that is, the
compound tones, individual partials, or combinational tones contained in them or resulting from
them, reinforce or enfeeble each other.   The tones
then do not coexist undisturbed in the ear ; they
mutually check each other's uniform flow."

Remarks on the Theory of Beats.

Of course it must be apparent that any dissonance
between two prime tones will be more distinctly
perceived, and produce a more direct effect upon
the ear than beats—*i.e.*, throbs or disturbances
—which   occur   between   upper   partial   tones.
Nevertheless, that these *do* impinge upon the sense
of audition, and *are* indications of a natural law
which enforces the just relation of combined tones,
there is ample evidence to prove.   In fact, all
well-informed musicians recognise the existence of
beats amongst the partials ;   and this is far more
distinctly, if not invariably, made manifest when
combined sounds are not related to each other by
simple ratios.

As we have before shown, beats occurring in
decidedly unrelated tones must be, of necessity, repeated with fresh effect in their partials, consequently
the unpleasing effect of dissonance in prime tones

becomes strengthened by their continued action in disturbances amongst the partials, and thus is added another reason for assuming that physical effects upon the ear precede psychical effects upon the mind, and that the coalescence of unrelated or dissonant tones upon the sense of audition is one of the chief reasons why the infraction of strict tonal laws produces an involuntary and perhaps unconscious sense of repulsion in the listener, whilst the converse of this position is inevitable, for obedience to natural law not only evolves the beautiful in sound, but also the recognition of what *is* beautiful in the satisfaction it conveys to the sense of audition.

It may be argued that if disturbing beats are liable to be felt by the ear as painful dissonances, when they occur amongst the partials of two fairly related prime tones, then the whole range of partials must result in forming dissonances after the octave or second partial has been reached. At this point it is necessary to recur to the theory of combinational tones before treated of. In fact, were it not reasonable to believe that vibrational tones might be thus fused into new and pleasing combinations, it would be difficult to account for the delight with which the ear listens to harmonious masses of sound, or how part music could become endurable at all.

Notwithstanding the fact that beats *do* occur, and

must occur, even in the extreme upper partials of all polyphonic music not limited to unison, we may be satisfied to accept the conclusions which arise from the study of combinational tones, and take for granted that, among well-related prime tones, the vibrations seldom if ever produce those painful sensations which grow out of dissonances or the interferences called beats.

To reinforce our position by further examples, we may say, when two notes of nearly the same pitch, such as C and C♯, are sounded together, the ear is not only made aware of the dissonance of the prime tones instantly, but, if moderately sensitive, can readily detect the interruption to the continuous flow of sounds which occur in the harmonics, careful attention to which will at once prove the action of beats.

On the other hand, when a prime and its perfect fifth are sounded together, a jar, or beat, can only occur between the fourth partial of the prime and the third partial of its fifth. Even then the question may arise whether a fusion of air waves may not dissipate the effect of dissonance and resolve the unrelated partials into a satisfactory combinational tone, the primes being related by simple ratios.

There are many methods by which beats may be detected, but it may be noted, in further illustration of this subject, that beats are most readily observed when they occur slowly. To determine their action, careful experiments must be conducted, both with slow and rapid interferences of this nature. It is true that beats occurring very rapidly, may not impress the ear with the same sense of dissonance as when they occur slowly. Nevertheless, although the ear may not be able to detect them, the confusion which the recurrence of many rapid impulses creates, makes a marked difference in the flow of continuous and uninterrupted sound.

Beats may be produced by way of experiment upon any musical instrument, but their best exemplification is obtained through tuning-forks or stopped organ pipes. With two tuning-forks of exactly the same pitch, the experiment can be conducted by sticking a little wax on the end of one. Then after striking them both, and applying them to the ear, or placing them on any wooden surface that will form a sounding board, the beats between the vibrations of the two instruments will be distinctly perceived. To produce beats between two stopped pipes sounding exactly the same pitch, flatten the tone of one of them, by inserting a finger slowly within the tube. Beats of compound tones can also be detected by striking a note on a pianoforte out of tune, when the difference between the two strings which serve the one note will give the unmistakable effect of a beat.

As we have referred at some length to the theories of Helmholtz on the subject of beats, it seems necessary to consider briefly what positions would arise from a full acceptance of those theories; also to observe what modifications are set up by Helmholtz himself to his own theories; a subject hereafter to be treated of, under the title of " Coincidences." To recur, however, more immediately to the propositions already laid down, we must review the basic principle enunciated by Helmholtz, namely : *That the beautiful in sound is the result of a physical effect produced on the sense of audition, and from thence impressing the mind agreeably; such physical effects as are calculated to coalesce with the ear and produce a sense of gratification to the mind being attainable only through the flow of a continuous or uninterrupted sound.*

As the subject under present consideration—that is to say, that beats are a set of special impulses which tend to break or altogether destroy the continuous and uninterrupted flow of sound—the importance which Helmholtz attaches to his theories concerning beats may be readily apprehended. Unfortunately for the practical application of those theories, however, all combined sounds—except those of exactly similar pitch—must generate in one or more of their upper partials the very phenomena which render the continuity of sound impossible. It follows, therefore, that to obtain that unbroken continuity of sound for which Helmholtz contends, we should either be limited to the combinations above indicated—melodic progression in which no combinations occur, except, as with ratios 1 : 1—or some modified form of combination, in which beats were either entirely counteracted or rendered inaudible. It is to this latter hypothesis that Helmholtz points the way in the theory of combinational sounds already treated of, and extended into still more plausible assumptions, in what he terms " coincidences "—a phase of tonal coalescence which, as he claims, may overpower and neutralise the action of beats, even if it does not actually arrest them.

Before advancing to the consideration of the theory of *coincidences*, however, we would summarise those positions on the subject of beats that have already been laid before the reader.

If it be true, as Helmholtz affirms, that the discovery of beats as an agent in producing intermittency of sound may tend to correct aberrations in the introduction of dissonances, either in the tuning of instruments or the combination of ill-related intervals, then the study of beats, as a

P

means of defining the laws by which the beautiful
in sound no less than the disturbing effects which
produce opposite results, must be recognised as an
important and necessary branch of musical science.
It must be admitted that there is much that is
merely hypothetical in a study that involves "high
harmonic attenuations" of sound, such as those
wherein the action of beats is assumed to occur in
the upper partials of combined prime tones. A
vast and exhaustive array of experiments proves,
however, that such disturbances *do* occur.

The theory that the beautiful in sound results
from an unbroken flow of continuous sound, hovers
on the verge of scientific demonstration, for whilst the
complete establishment of these occult principles of
tonal law may yet be considered as held in abeyance,
enough has been shown to prove that these tonal
laws may and MUST gain by such suggestions as
lead to careful analysis, patient experiment, and a
study of plausible hypotheses, based upon demon-
strable proofs of acoustical principles in action.
Hence—as above affirmed—the analysis of beats,
and all the attendant sources of modification, which
may tend to prevent their occurrence or quench
their action, should be regarded as a subject
of primary importance with the musician.

Chapter VI

ON COINCIDENCES.

B Y the term "coincidence" in music is implied
the agreement in pitch of two sounds.

Helmholtz's theory of coincidences refers to the
coincidences which occur between sounds not wider
apart than two octaves.

Coincidences with the above-stated limitations
may occur between

1. Primes.

2. The upper partials of one prime and the
upper partials of another.

3. A prime and an upper partial of another prime.

4. A differential tone and a generating tone.

5. Differential tones.

6. A summational tone and a partial of a generating tone.

Coincidences occur between primes when a sound
is combined with its unison, in which case, all the
partials of each prime coincide. Thus it will be

seen that if the two primes are C and C the partials
are as follows :—

Partial 8	c″	┌─────────┐	c″
„ 7	bb′	┌────────┐	bb′
„ 6	g′	┌────────┐	g′
„ 5	e′	┌───────┐	e′
„ 4	c′	┌───────┐	c′
„ 3	g	┌──────┐	g
„ 2	c	┌─────┐	c
„ 1	C	┌────┐	* C

There is coincidence between a prime and a
partial when one prime is combined with another,
in which the ratio of the pitch number is either
$1 : 2$, $1 : 3$, or $1 : 4$. Thus, if C be combined with
its octave c there is coincidence between c and the
second partial of C.

When C is combined with its 12th (g, $1 : 3$) there
is coincidence between g and the third partial of C

There is some coincidence between partials of
higher vibrational number when one prime is combined with another, and the ratio of pitch number
is either as $4 : 5$, $3 : 5$, $5 : 6$, or $5 : 8$.

Helmholtz carries the theory of coincidences
still farther than the agreement or harmonization
of certain partials amongst themselves, for he
argues that it may be—in fact must be—that
coincidences will counteract the influence of
beats. He says: "There are certain combi-

---

* The line ┌─────┐ shows the coincidence.

nations between compound vibrations which, though accompanied by beats, are yet so powerfully affected by coincidences amongst the partials that the actual harmony of the united sounds can be perceived. The special ratios by which the generating tones, as well as the upper partials, resist beats are those of the fifth, fourth, major third and sixth, and the minor third and sixth.

The coincidence in the case of a sound (say C) combined with its fifth, overpowers the disturbing influence of the fourth and fifth partial of the one prime with the third partial of the other.

ILLUSTRATION.

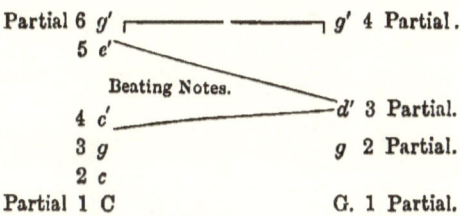

Partial 6 g' g' 4 Partial.
5 e'
Beating Notes.
4 c' d' 3 Partial.
3 g g 2 Partial.
2 c
Partial 1 C   G. 1 Partial.

When a sound (say C) is combined with its fourth, the coincidence of the fourth partial of C, with the third partial of F overpowers the beats between the third partial of one prime and the second partial of the other.

ILLUSTRATIONS.

Coincidence.

c' c'
g Beating Notes. f
c
C   F

When a sound (say C) is combined with its major third, the coincidence of the fourth partial of E with the fifth partial of C overpowers the beats sufficiently to prevent interference of sound by beats.

```
Partial  5  e' ┌───────────┐ e'  4  Partial.

         4  c'   Beating Notes
         3  g                    b   3  Partial.
         2  c                    e   2  Partial.
Partial  1  C                    E   1'  Partial.
```

When a sound is combined with its major sixth (C with A) the coincidence of the fifth partial of one prime with the third partial of the other overpowers the interference of beats sufficiently to preserve the continuity of sound.

```
         5  e' ┌───────────┐ e'  3  Partial.
         4  c'
         3  g    Beating Notes   a   2  Partial.
         2  c
Partial  1  C                    A   1  Partial.
```

When a sound is combined with its minor sixth (C A♭) the eighth partial of one prime coincides sufficiently with the fifth partial of the other to preserve continuity of sound.

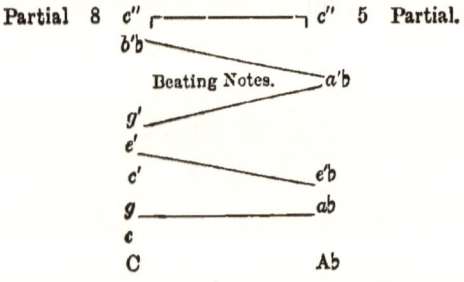

```
Partial  8  c'' ┌───────────┐ c''  5  Partial.
            b'b
                 Beating Notes.    a'b
            g'
            e'
            c'                     e'b
            g                      ab
            c
            C                      Ab
```

It will be seen, according to the definitions of Helmholtz, and the importance which he attaches to coincidences, as factors in the modification of such disturbances as interfere with the continuity of sound, that he ranges these coincidences into four groups, viz., the absolute, the perfect, the medial, and the imperfect.

" The first or absolute kind requires that the prime tone of one of the combined tones shall coincide with some partial of the other."

" *Examples.*—The octave, twelfth, and double octave."

" The second, or perfect kind, requires that one of the *lower** partials of the prime tone shall coincide with *one* of the *lower* partials of the second prime tone."

" *Examples.*—The fifth and the fourth."

" The third or medial kind requires a coincidence between one of the upper partials of each of the two prime tones."

" *Examples.*—The major third and the major sixth."

" The fourth or imperfect kind requires that there be some coincidence among the partials, even though it be sought for amongst the extreme upper partials of the two primes."

---

* That is not far distant above the prime.

" *Examples.*—The minor third and the minor sixth."

Considering the graduated degrees by which the consonances of the first group merge into the dissonances allowed in the last, it seems not only difficult but unphilosophical to draw sharp lines of demarcation between the consonances and dissonances which occur in the attenuated realms of upper partials, provided always that the one grand mathematical principle of simple ratios between the generating tones and perfect tune between all the instruments employed be carefully carried out.

According to Helmholtz, however, though the sounds combined of ratios 2 : 3, 3 : 4, 4 : 5, 3 : 5, 5 : 6, and 5 : 8 produce beats in their vibrational action, coincidences amongst some of the partials, tend to overpower the intermittency of sound produced by those beats, and confirm the theory that the combination of tones, based upon the law of simple ratios, *does* tend to create harmony, even amidst the high harmonic attenuations of upper partial vibrations. The theory of differential tones suggests the same tendency to correct the disturbing influence of beats, for whilst Helmholtz argues that beats are created by combinational tones, he admits that this is only the case when the generating tones and upper partials produce beats.

To reiterate some of the positions laid down above, we might say that coincidences should arise, or in other words, that consonances, defined by simple ratios, should exist, between the combined tones illustrated in the above groups. Allowing that beats are disturbances in that continuity of sound so necessary to the beautiful in tone, that it is affected by the slightest jar which may produce dissonance even amongst the upper partials, it may be questioned, whether two primes of different pitch being sounded simultaneously can produce coincidences between third and fourth or more widely removed upper partial vibrations? Helmholtz's theory affirms that such coincidences *can* and *do* neutralise the action of beats. So far this theory may be accepted as satisfactory. But even if it were considered as liable to objections, unnecessary at present to advance, the various modifications suggested by the additional theories of combinational tones, open up possibilities in the realm of tonal science which cannot be set down as finally settled or closed against future researches in the direction of vibratory action.

In fine, every candid mind must admit that the problems of tonal science are not yet solved.

Coalescence of air particles may be found to follow the same law by which the larger rain drops on the window pane swallow up the smaller, and thus the earlier and more powerful momentum of the lowest upper partials may overpower the finer and more attenuated tones of the ascending later upper partials. Combined upper partials and combinational tones point to the possibility of other probable coalescences, and analogies may yet be found to exist · between the action of chemical affinities in atoms and those of air waves.

Allowing that none of the theories concerning vibratory tones above treated of are free from objections, they are still suggestive, and all tend to the conclusion that the pleasurable emotions which arise from listening to purely harmonious musical sounds are derived from immutable principles of law and mathematical proportions of tonal arrangement.

CHAPTER VII

CONCLUSION.

NOTWITHSTANDING the wide fields of exact
science which the musicians of different ages
have traversed, and the wealth of discovery concerning the natural principles of tone which have
yielded up their secrets of art during the present
century, there are still a vast number of persons
who, having some practical knowledge of music,
contend that its matchless charms are derived only
from the capacity of the mind to appreciate them,
or, in other words, that music is a matter of taste,
sentiment, and educational culture.

To assert this, in view of the discoveries effected
in musical science by modern research, would be as
fallacious as to refer the sensations of heat and cold
to taste, sentiment, and culture.

Whilst we *know* that the relations subsisting
between the atmosphere and the human system
originate the sensations of heat and cold, and that
the degree in which those sensations prevail may be
modified by personal idiosyncrasies, these latter are
modifications only, not causes. Even so may we
determine, that it is the arrangement of tones,

whether in the order of natural law or in violation thereof, in which originate the sensations of pleasure or pain which musical sounds produce upon the nerves of audition.

In matters of detail all the theories touched upon in the preceding pages have yet to be outwrought to far more practical forms of demonstration than they have at present attained.

The basic principles from which the beautiful in sound can alone be evolved, however, are being too clearly apprehended by musical scientists to admit of dependence on mere psychological feeling or ideality.

Still, as an example of the objections to which even the most plausible theory is open, unless it can be proven by well-established demonstrations, we may call attention to the fact that, whilst Helmholtz satisfactorily enough accounts, in his highly-elaborate theoretical methods, for the harmonious character of the fourth by making its third partial coincide with the fourth partial of the prime, and in more than one instance has cited the "coincidences" between different degrees of the "partials" as sufficient to counteract the effect of disturbing dissonances, the weak point in his argument is, when he attempts to measure any of the higher and weaker partials against the momentum and force of the lower ones. If the loudness of the vibratory tone diminishes in proportion to its distance from the generating tone, until, after reaching the sixth interval, it seems to struggle up, as it were, to a minor seventh, and fades off into the finest attenuations of tone, scaling bold intervals no longer, but ascending the next three notes by successive tones, these results give colour to the inference that the power would not be sufficient to interrupt the lower and more distinctive tones in the form of "beats," or even to reach the unaided ear with a sense of complete dissonance.

The laws of motion which apply to projectiles operate with variations of force and speed upon impulses of any kind communicated to the air, including the vibratory action of musical tones.

Whilst the knowledge we have attained to concerning tonal laws, may be considered merely rudimental, compared with the vast fields which lie open for future research, so we are scarcely justified as yet in attempting to educe arbitrary rules of tonal motion from high attenuations of sound amongst partials. "It may be so" is the only safe dictum we can at present apply to vibratory effects, until we have still more fully gauged the untrodden realms of law, which the discovery of "Harmonics" opens up to us.

Three very marked and important steps in the progress of musical science have undoubtedly been gained by the researches of the last two centuries.

The most important of these is the realization, that the true seat of perception concerning that which is pleasing or repulsive to the mind in musical sound, is derived from a coalescence between the waves of air set in motion by music, and the human ear, or in other words, a direct physical relation between that sound and the auditory nerves.

In this respect, it is not enough to say, the sense of hearing informs the mind. The sense of hearing bears some subtle but mathematical relation to *quality* in sounds, so that one class of sounds inpinges on the auditory nerves with sensations of pain, and another with its reverse. The real definitions concerning these wonderfully occult relations between the anatomical structure of the ear, and the order of vibrations which sounds produce in the air, is amongst the disclosures *that shall be*, in the near future, as yet they are not fully attained to.

Secondly, we have learned, that the musical tones which impinge upon the auditory nerves with a sense of pleasure, are those which conform in their arrangement to certain mathematical proportions to be observed between the pitch of tones which mark intervals, whether in succession or combination.

Thirdly, we have proved, that the intervals between the pitch of different tones—whether in succession or combination—must be determined by simple ratios, or whole numbers, if we would produce such an arrangement as would affect the ear pleasurably.

The whole subject of " harmonics, combinational tones, beats, and coincidences," forms, as must be seen, an added chapter which really grows out of the experiments which modern science has pursued, in order to demonstrate the three primary principles laid down in the above propositions.

In the realm of art, it is of no use to question why certain effects inevitably follow certain causes, or refuse to accept results, because we cannot satisfactorily explain them. *They exist*, and will yet yield up their secrets to intelligent effort and patient investigation. Such is the experience of true philosophy the world over, and the centuries through.

We know that certain forms and colours affect the mind pleasurably or the reverse, and modern research tends to show that these mental sensations grow out of physical causes, and that nature's laws, not the human eye, define what is beautiful in form and colour.

The eye only perceives, and the mind appreciates according to its organic capacity to do so; the law of the beautiful remains intact in nature, let who will apprehend it or not. Precisely the same is the law of the beautiful in sound. The principles are fixed and arbitrary. Their discovery and application is the source of delight to the mind that can appreciate them, their infraction is the cause of pain and repulsion to the sensitive ear. The more thoroughly we guage the sources of all the effects which appeal to the senses, the more surely do we find they grow out of inherent principles written in the order of the universe, while the perceptions of those principles, in the highest degrees of mental culture, are recognized by the sensations of pain or pleasure they occasion.

It was an immense triumph for the Greeks to discover more than two thousand years ago that the sensations of pain or pleasure, evoked by musical sounds, resulted from, or conformity to, infraction of arbitrary tonal laws. It was a vast stretch of the antique mind into the realm of universal order, to apprehend that intervals between sounds of different pitch should be determined by simple ratios. Many causes conspired to hinder practical application of the knowledge which Greek philosophy had discovered, but the perception of these inherent principles in nature, only illustrates the axiom that " truths are eternal," though the methods of applying them are dependent upon the logic of events, time, and change. Thus the more extended branches of science which modern research has revealed, enables us to supplement Greek philosophy on the nature of tonal law at the point where that philosophy stopped short. The discoveries of Galileo concerning the nature of oscillating bodies, necessitated the construction of instruments of measurement. Instruments thus fashioned must be improved upon, and so, multiplied themselves; and these again have promoted fresh researches, which have culminated in the discovery of "harmonics," "combinational tones," coincidences," and other occult principles which underly the action of musical tones and vibrating bodies. And thus the truths that were partially revealed in Greek philosophy have been vindicated, whilst its errors and shortcomings have been fully supplemented. How much more remains for futurity to achieve, the highest flights of transcendentalism may vaguely dream of, but can never attempt to limit. Whilst we may felicitate ourselves, with full justice, on what we have at present gained, and hope everything from the future, there is one feature, in the development of musical science which is anything but a subject of self-

Q

gratulation; and this is the fact, that musical progress has proceeded—in the middle ages at least—with far less certainty and speed than any other of the sister arts. Even now, and despite the widespread opportunities for the cultivation of musical science which the nineteenth century affords, the opinion of the majority on the subject of the beautiful in sound, still inclines to the belief that it is chiefly a matter of taste, feeling, and sentiment, with a certain amount of allowance for the influences of civilization and culture.

Something of this widespread ignorance of elementary knowledge on the science of music, proceeds doubtless from the multiplication of cheap and superficial methods of training—from the exaltation of mechanical execution over the understanding of the principles of art—and perhaps also, from the effect of the long hiatus of twenty centuries which has intervened since Grecian sages enunciated the basic principles of musical art, and modern research has been able to apply them practically.

Language would fail to depict the universality of the divine art of music, or do justice to the deep and imperishable hold which it has maintained and ever must continue to exert over the mentality of the human race.

And yet, as before stated, all the sister arts seem to have been far more readily ranged under the domain of natural law, and practised by the rules and observances of science, than music.

Architecture has, ages ago, embodied the profoundest rules of geometry, and painting has achieved even grander triumphs in the middle ages than in modern times.

Sculpture, likewise owes more to the masters of classic and mediæval art than to those of our own period, and when we proceed to analyze the sources of superiority in all these branches of art, we unhesitatingly affirm that they consist of fidelity to the fundamental principles of nature.

For example, what should we think of any artist who, under the influence of taste, fancy, or sentiment, would paint the sun green, or the grass yellow? Of the sculptor who thought proper to embody his own peculiar views of the beautiful by carving a giant's hand on an infant's form, or the architect who, to please his own ideality, would have upreared a colossal tower upon a cottage foundation?

In these, and all other departments of art, nature has proved to be the only true, faithful, and authoritative model, and nature's laws—not the wild fantasies of the human mind—will ever be the final arbiter of what is false or true in art.

Isolated fragments of beauty are valueless, unless they maintain just relations to each other in organism, and preserve the most perfect harmony of relation between their several parts.

The one single sound of a church bell means death ; annihilation of present realities. A peal of bells of different though well-related tones means life, joy, action, and all the possibilities that life and action imply. The simple sound of the tolling bell may take any pitch. The succession of intervals and differences of pitch in the peal, must be duly related by mathematical proportions, or their chime is unendurable.

And thus, unless the principles of just relation among the parts, which apply to other arts, are incorporated in tonal science, music is not only an anomaly amongst the arts, but it becomes like the tolling bell, the symbol of death, and needs the resurrecting principle of organic science in number, order, and variety, to awaken it into life and meaning.

Æsthetics, the science of the beautiful, and mathematics, the basis of all science, should walk hand in hand in music as in all other arts. In fact, law should be recognised as the corner stone of the structure on which æsthetics rest.

The Greek mathematicians, in searching for the law, forgot to consult the taste which craves for the beautiful, whilst Aristoxenes and his followers, in catering only for the taste, reared up a superstructure which lacked the basis of fundamental principles to rest upon.

The aim of modern philosophy must be to place the beautiful in sound on the substructure of pure science, and not until music takes this high and unassailable position can its true genius be appreciated, or its noblest purposes fulfilled.

At present we are only in the path of discovery, not in the field of actual achievement.

Still we begin to realize the goal we must reach, and our imperfect discernment of great truths cannot mask their significance when they all point one way.

Dimly perceived by the intuitions of the meta-physical Greeks, almost captured by the eighteenth century mathematicians, and determinately grappled with by the musicians of the present day, patient research and continued experiment cannot fail to bring to the noonday light of scientific demonstration THE FACT that, MUSIC IS THE SPEECH OF NATURE, and that just as we arrange her beautiful harmonies in accordance with her own immutable laws, so surely shall we realize the import of that speech in the pleasure or pain which musical tones communicate to the sense of audition.

Taste, sentiment, and the whole range of mental emotions, do, and often will, confuse just perceptions in matters of art where personal judgment is to be given, but laws and principles remain unalterable however long they may have to wait for recognition. Meantime, education, culture, and æsthetical feeling must all be admitted as factors for clear mental perception of the beautiful in sound.

It is one thing to know the law, quite another to realize its working in individual consciousness, especially when the sense of audition—which is as different in different organisms as are physiological varieties of the one structure—is admitted to be the arbiter of what is most pleasing or otherwise in musical tones.

Still the result of modern, as well as ancient philosophy, assures us that the highest scope and power of which music is capable, will never be ultimated, until fertile fancy and inventive genius are allied to mathematical proportions, and regulated by the fundamental principles of nature.

Then will the erratic flights of fancy be reduced to the graceful order of pure harmonies, and the grand inspirations of musical ideality . cease to be marred by the vague speculations upon what music might be, or, still more likely, what it will never be.

What music *ought to be*, and ultimately *must* be, is now beginning to dawn upon the mental perception of the true scientist The day may not be far distant then, when, instead of regarding with contemptuous incredulity the theories of Greek sages, concerning the analogies of music with the motions and tones of the universe, we may begin reverently to acknowledge that music and its wonderful array of occult powers, are the best and only interpreters of nature's sublime meanings, and the best and most universal illustration of Pope's deeply philosophic couplet,

"All are but parts of one stupendous whole,
Whose body Nature is, and God the soul."

FINIS.

INDEX.

INDEX.

JOHN HEYWOOD, Excelsior Printing and Bookbinding Works, Manchester.